# Apollo 12

## The NASA Mission Reports
## Volume 2

Compiled from the archives & edited
by Robert Godwin

Special thanks to:
Nadine Andreassen
Benny Cheney
Colin Fries
Steve Garber
John Hargenrader
Margaret Persinger
Kipp Teague


We acknowledge the financial support of the Government of Canada through the
Book Publishing Industry Development Program for our publishing activities.
Published by Apogee Books an imprint of Collector's Guide Publishing Inc., Box 62034, Burlington, Ontario, Canada, L7R 4K2
Printed and bound in Canada
Apollo 12 - The NASA Mission Reports Volume 2
by Robert Godwin
ISBN 1-894959-16-7

# Apollo 12

## The NASA Mission Reports
## Volume 2

(from the archives of the National Aeronautics and Space Administration)

## INTRODUCTION

It's 35 years since the Intrepid voyage to the Ocean of Storms. Ask the average person to name the crew and you'll be lucky if they can name even one of them. Ironically this collective amnesia is likely to be transitory, in a couple of hundred years from now the names of Conrad, Gordon and Bean will probably find their true place in the annals of human history. It is my sincere hope and fervent belief that by the 23rd century we will be a true space-faring civilization, one which remembers its forebears with the honor and respect they earned.

Meanwhile, back here in the turgid backwaters of the early 21st century some of us do what we can to keep the flame of hope alive. The average person may look into the following pages and see only statistics and charts but if you look hard enough you can see that they contain the hard-earned kernels of our collective future. Lest this data languish into obscurity we have replicated it here for future students to ponder.

Captured here are the details of leaky spacesuits, rockets struck by lightning, spacesuit cooling systems which stop working because the spacecraft's door got stuck closed and cameras which were disabled by dust. Then there was the hydrogen tank which failed at the last minute and was fixed by pulling its counterpart from Apollo 13, or the outrageous pinpoint landing by Pete Conrad through a pall of lunar dust, so bad he couldn't see the surface from 40 feet up. If you want to know why that TV camera failed or just how badly decomposed the robotic Surveyor had become after three years on the moon, it's all here.

Let's also not forget the fact that Apollo 12 was the first to use the hybrid trajectory with its so-called non-free-return. Such a trajectory for Apollo 13, without correction, would have been fatal.

Then as if to really test the mettle of its hardy crew, the Command Module Yankee Clipper threw one last curve-ball when the drogue parachute deployed almost 20 seconds late—just imagine how that 20 seconds must have felt to the crew and controllers—before the Command Module hit the water at a punishing 15g's.

The photographic record of Apollo 12 is, to put it politely, sketchy. The TV camera failed during EVA, the 16mm movie camera got so much dust on it that much of the footage from the LM window is blurred and unwatchable before it finally jammed during lunar lift-off. When it was working, for much of the time, nothing was happening in the picture. The chest-mounted Hasselblad became clogged to the point where the crew couldn't get it on and off their chest-mounts and as a final coup de grace the 16mm pounced from its mount during splashdown and attempted to spontaneously disassemble the lunar module pilot's head! Despite all of this there are many excellent still pictures and the moving picture record that does exist has its great moments. To try and do justice to this most eccentric mission I have attempted to accumulate some of the highlights on the accompanying DVD-Video disc. Some artistic liberties were taken, such as during the landing sequence the data reports the range to the proposed landing site, which was in fact 500 feet from the actual landing site, so during the last few seconds the range numbers are "fudged" somewhat. I figured people would rather see a big fat "0" at touchdown so it is only an approximation. The rendezvous and docking footage was also an interesting challenge to combine since the TV only kicks in half way through and then for some unfathomable reason drops-out momentarily, while Dick Gordon actually had the presence of mind to change the cartridge and the frame rate in the middle of the rendezvous. As if this wasn't hard enough, just finding the audio was unexpectedly difficult. Regardless, I hope you enjoy it and join me in tipping your hats to Pete, Al and Dick. Go Navy!

Rob Godwin (Editor)

## CONTENTS

MSC-018

NATIONAL AERONAUTICS AND SPACE ADMINISTRATION

# APOLLO 12 MISSION REPORT

MANNED SPACECRAFT CENTER
HOUSTON, TEXAS
MARCH 1970

# MSC-018

NATIONAL AERONAUTICS AND SPACE ADMINISTRATION
APOLLO 12 MISSION REPORT
MANNED SPACECRAFT CENTRE
HOUSTON, TEXAS
MARCH 1970

| Mission | Spacecraft | Description | Launch Date | Launch Site |
|---|---|---|---|---|
| PA-1 | BP-6 | First pad abort | Nov. 7, 1963 | White Sands Missile Range, N. Mex |
| A-001 | BP-12 | Transonic abort | May 13, 1964 | White Sands Missile Range, N. Mex. |
| AS-101 | BP-13 | Nominal launch and exit environment | May 28, 1964 | Cape Kennedy, Fla. |
| AS-102 | BP-15 | Nominal launch and exit environment | Sept. 18, 1964 | Cape Kennedy, Fla. |
| A-002 | BP-23 | Maximum dynamic pressure abort | Dec. 8, 1964 | White Sands Missile Range, N. Mex. |
| AS-103 | BP-16 | Micrometeoroid experiment | Feb. 16, 1965 | Cape Kennedy, Fla. |
| A-003 | BP-22 | Low-altitude abort (planned high-altitude abort) | May 19, 1965 | White Sands Missile Range, N. Mex. |
| AS-104 | BP-26 | Micrometeoroid experiment and service module RCS launch Environment | May 25, 1965 | Cape Kennedy, Fla. |
| PA-2 | BP-23A | Second pad abort | June 29, 1965 | WhiteSands Missile Range, N. Mex. |
| AS-105 | BP-9A | Micrometeoroid experiment and Service module RCS launch Environment | July 30, 1965 | Cape Kennedy, Fla. |
| A-004 | SC-002 | Power-on tumbling boundary abort | Jan. 20, 1966 | White Sands Missile Range, N. Mex. |
| AS-201 | SC-009 | Supercircular Entry with high heat rate | Deb. 26, 1966 | Cape Kennedy, Fla. |
| AS-202 | SC-011 | Supercircular Entry with high heat load | Aug. 25, 1966 | Cape Kennedy, Fla. |

# APOLLO 12 MISSION REPORT

*PREPARED BY*

*Mission Evaluation Team*
*APPROVED BY*

*James A. McDivitt*
*Colonel, USAF*
*Manager, Apollo Spacecraft Program*

*NATIONAL AERONAUTICS AND SPACE ADMINISTRATION*
*MANNED SPACECRAFT CENTER*
*HOUSTON, TEXAS*
*March 197*

Apollo 12 Lift Off

# 1.0 SUMMARY

The Apollo 12 mission provided a wealth of scientific information in this significant step of detailed lunar exploration. The emplaced experiments, with an expected equipment operation time of 1 year, will enable scientific observations of the lunar surface environment and determination of structural perturbations. This mission demonstrated the capability for a precision landing, a requirement for proceeding to more specific and rougher lunar surface locations having particular scientific interest.

The space vehicle, with a crew of Charles Conrad, Jr., Commander; Richard F. Gordon, Command Module Pilot; and Alan L. Bean, Lunar Module Pilot; was launched from Kennedy Scace Center, Florida, at 11:22:00 a.m. e.s.t. (16:22:00 G.m.t.) November 14, 1969. The activities during earthorbit checkout, translunar injection, and translunar coast were similar to those of Apollo 11, except for the special attention given to verifying all spacecraft systems as a result of lightning striking the space vehicle at 36.5 seconds and 52 seconds. A non-free-return translunar trajectory profile was used for the first time in the Apollo 12 mission.

The spacecraft was inserted into a 168.8- by 62.6-mile lunar orbit at about 83-1/2 hours. Two revolutions later a second maneuver was performed to achieve a 66.1- by 54.3-mile orbit. The initial checkout of lunar module systems during translunar coast and in lunar orbit was satisfactory. At about 104 hours, the Commander and the Lunar Module Pilot entered the lunar module to prepare for descent to the lunar surface.

The two spacecraft were undocked at about 108 hours, and descent orbit insertion was performed at approximately 109-1/2 hours. One hour later, a precision landing was accomplished using automatic guidance, with small manual corrections applied in the final phases of descent. The spacecraft touched down at 110:32:36 in the Ocean of Storms, with landing coordinates of 3.2 degrees south latitude and 23.4 degrees west longitude referenced to Surveyor III Site Map, First edition, dated January 1968. One of the objectives of the Apollo 12 mission was to achieve a precision landing near the Surveyor III spacecraft which had landed on April 20, 1967. The Apollo 12 landing point was 535 feet from the Surveyor III.

Three hours after landing, the crewmen began preparations for egress and egressed about 2 hours later. As the Commander descended to the surface, he deployed the modularized equipment stowage assembly, which permitted transmission of color television pictures. The television camera, however, was subsequently damaged. After the Lunar Module Pilot had descended to the surface and erected the solar wind composition foil, the crew deployed the Apollo lunar surface experiment package. On the return traverse, the crew collected a core-tube soil specimen and additional surface samples. Also, an Apollo erectable S-band antenna was deployed for the first time. The duration of the first extravehicular activity period was 4 hours.

Following a 7-hour rest period, the second extravehicular activity period began with preparation for the geology traverse. Documented samples, core-tube samples, trench-site samples, and gas-analysis samples were collected on the traverse to the Surveyor III spacecraft. The crew photographed and removed parts from the Surveyor. Following the return traverse, the solar wind composition foil was retrieved. The second extravehicular activity period lasted 3-3/4 hours. Crew mobility and portable life support system operation, as in Apollo 11, were excellent throughout the total 7-hour 46-minute extravehicular period. Approximately 74.7 pounds of lunar material were collected for return to earth, as well as the Surveyor parts.

The ascent stage lifted off the lunar surface at 142 hours. After nominal rendezvous sequence, the two spacecraft were docked at 145-1/2 hours. The ascent stage was jettisoned following crew transfer and was maneuvered by remote control to impact on the lunar surface; impact occurred at 150 hours approximately 40 miles from the Apollo 12 landing site.

After a period of extensive landmark tracking and photography, trans-earth injection was accomplished with the service propulsion engine at 172-1/2 hours. The lunar orbit photography was conducted using a 500-mm long-range lens to obtain mapping and training data for future missions

During transearth coast, two small midcourse corrections were executed, and the entry sequence was normal. The command module landed in the Pacific Ocean at 244-1/2 hours. The landing coordinates, as determined from the onboard computer, were 15 degrees 52 minutes south latitude and 165 degrees 10 minutes west longitude. After landing, precautions to avoid lunar organism back-contamination were employed. The crew, the lunar material samples, and the spacecraft were subsequently transported to the Lunar Receiving Laboratory.

## 2.0 INTRODUCTION

The Apollo 12 mission was the twelfth in a series of flights using Apollo flight hardware and was the second lunar landing. The purpose of the mission was to perform a precise lunar landing and to conduct a specific scientific exploration of a designated landing site in the Ocean of Storms.

Since the performance of the entire spacecraft was excellent, this report discusses only the systems performance that significantly differed from that of previous missions. Because they were unique to Apollo 12, the lunar surface experiments, the precision landing operation, and lunar dust contamination are reported in sections 3, 4, and 6, respectively.

A complete analysis of all flight data is not possible within the time allowed for preparation of this report. Therefore, report supplements will be published for certain Apollo 12 systems analyses, as shown in appendix E. This appendix also lists the current status of all Apollo mission supplements, either published or in preparation. Other supplements will be published as the need is identified.

In this report, all actual times prior to earth landing are elapsed time from range zero, established as the integral second before lift-off. Range zero for this mission was 16:22:00 G.m.t., November 14, 1969. Greenwich mean time is used for all times after earth landing as well as for the discussions of the experiments left on the lunar surface. All references to mileage distance are in nautical miles.

## 3.0 LUNAR SURFACE EXPLORATION

This section contains a discussion of the formal experiments conducted for Apollo 12 and presents a preliminary laboratory assessment of returned samples. The experiments discussed includes those associated with the Apollo lunar surface experiments package and the solar wind composition, lunar geology, lunar surface photography, and multispectral photography experiments. The evaluations in this section are based on the data received during the first lunar day. All final experiment results will be published in a separate science report when the detailed analyses are complete (appendix E).

Lunar surface scientific activities were performed essentially as planned within the allotted time periods. Three hours after landing, the crew began preparations for egress and the first traverse of the lunar surface. During the first extravehicular activity period, which lasted 4 hours, the crew accomplished the following:

a. Deployed the modularized equipment stowage assembly, which permitted transmission of color television pictures of the Commander descending the lunar module ladder

b. Transferred a contingency surface sample to the lunar module

c. Erected the solar wind composition foil

d. Collected a core-tube soil specimen and additional surface samples

e. Deployed the Apollo lunar surface experiments package for an extended collection of lunar scientific data via a radio link.

The experiments package included a cold cathode gage, a lunar surface magnetometer, a passive seismometer, a solar wind spectrometer, a dust detector, and a suprathermal ion detector. A brief description of the experiment equipment is presented in appendix A. Certain difficulties in deploying the equipment are mentioned in this section and are discussed in greater detail in section 14.3. Anomalies in the operation of the equipment since activation are also mentioned, but the nature and cause of each experiment anomaly will be summarized in a later science report (appendix E).

Following a 7-hour rest period, the second extravehicular activity period began with preparations for the geology traverse. The duration of the second extravehicular activity was 3-3/4 hours, during which the crew accomplished the following:

a. Collected documented, core-tube, trench-site, and gas-analysis samples.

b. Photographed the Surveyor III and retrieved from it a cable, a painted tube, an unpainted tube, the television camera, and the scoop.

c. Retrieved the solar wind composition foil.

Crew mobility and perceptibility, as in Apollo 11, were excellent throughout both extravehicular periods. The discussion in the following paragraphs is based largely on real-time information and crew comments.

### 3.1 APOLLO LUNAR SURFACE EXPERIMENTS PACKAGE

The Apollo lunar surface experiments package was deployed on the lunar surface at 116 hours (fig. 3-1), and the experiments were activated between 118 and 124 hours. After the initial difficulty in removing the radioisotope fuel capsule from its transporting cask (see section 14.3.3), the crew installed the capsule in the radioisotope thermoelectric generator. The experiment package transmitter was turned on by ground command approximately 69 minutes after the fueling of the generator. At the time of activation the power output of the radioisotope thermoelectric generator was 56.7 watts; as the generator warmed up, the power output steadily increased to 73.69 watts and has remained nearly constant at that level.

Figure 3-1 Lunar Module Pilot lifting Apollo lunar surface experiments package prior to deployment traverse.

The transmitter downlink signal strength was minus 139 dBm at the time of activation and has remained constant at about minus 140 dBm. The execution of uplink commands verified normal communications. Several commands have not shown command verification in telemetry data but were verified by functional changes in the experiment operation. The overall performance of the central station, shown in figure 3-2, has been exceptionally stable. Temperatures at various locations on the thermal plate, which supports electronic equipment, are shown in figure 3-3, and the average thermal plate temperatures have been well within the excected maximum values since activation.

Discussions of the preliminary performance and, when available, scientific results for each of the studies in the experiment package are presented in the following paragraphs.

### 3.1.1 Dust Detector

Output data from the dust detector cells are shown in Figure 3-4. All readings are close to expected values and show no evidence of natural dust accumulations. An increase in the cell 2 output was seen at lunar module lift-off. Data from cell 2 show that the sun incidence angle was normal to the cell face about 6 hours prior to actual lunar noon, indicating the package is probably tipped about 3 degrees to the east.

Figure 3-2 Central power station cables and flat-tape power.

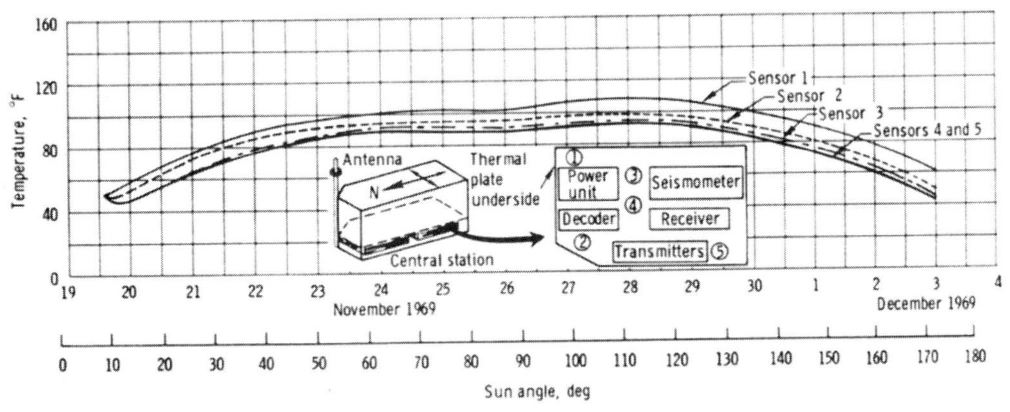


Figure 3-3. - Central station thermal plate temperatures.

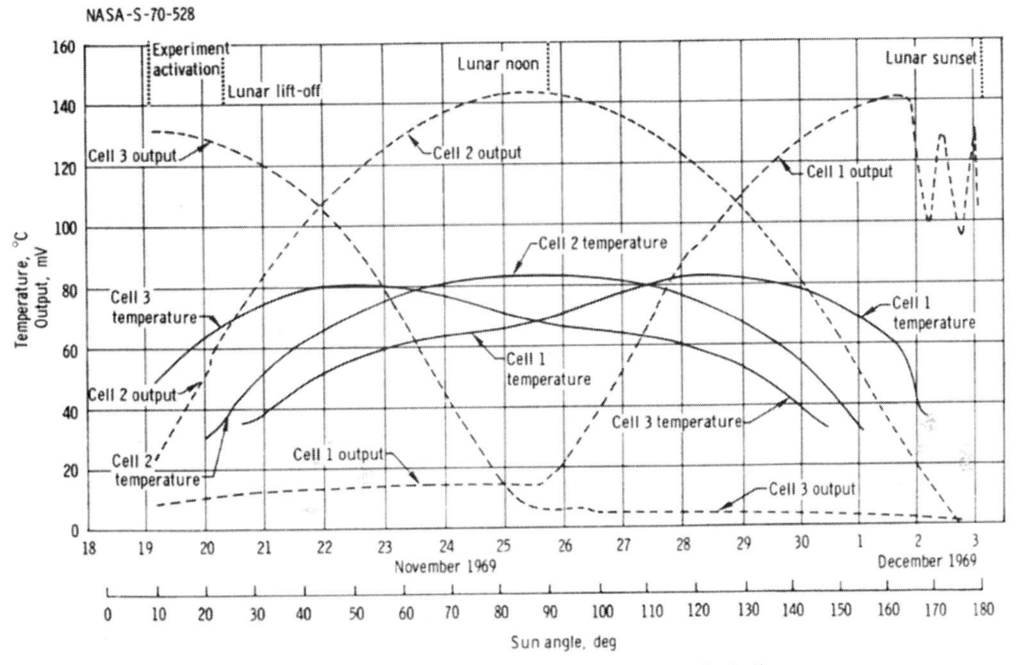


Figure 3-4. - Dust detector data during first lunar day of activation.

### 3.1.2 Passive Seismometer Experiment

The passive seismic experiment, shown in figure 3-5, has operated as planned with the exceptions noted. The sensor was installed at a location west-northwest from the lunar module (fig. 3-6) at a distance of 130 meters from the nearest footpad. The crew reported that tamping the surface material with their boots was not an effective means of preparing the surface for emplacement because the degree of compaction is small. Spreading the thermal shroud over the surface was difficult, because in the lunar gravity, the lightweight Mylar sheets of this shroud would not lie flat (see section 14.3.4) .

Instrument performance.- The passive seismic experiment has operated successfully since activation; however, instrumentation difficulties have been observed.

The short-period vertical-component seismometer is operating at a reduced gain and fails to respond to calibration pulses. Detailed comparisons between signals observed on both the long- and short-period vertical-component seismometers has led to the initial conclusion that the inertial mass of the short-period seismometer is rubbing slightly on its frame. Nominal response is observed for signals large enough to produce inertial forces on the suspended mass which apparently exceed restraining frictional forces. The threshold ground-motion acceleration required to produce an observable signal cannot be determined accurately, but it is probably less than $8 \times 10^{-4}$ cm/sec$^{2}$ , which corresponds to surface motions of 2 millimicrons at a frequency of 10 hertz. On December 2, 1969, a series of square-wave pulses were observed on the short-period vertical trace over a period of approximately 13 hours. The pulse amplitude was constant and was approximately equal to a shift in the third least-significant bit of a telemetry data word. These pulses are also observable on the records from the long-period seismometers, but with reduced amplitude. The problem is believed to be in either the analog-to-digital converter or the converter reference voltage.

The response of the long-period vertical seismometer to a calibration pulse was observed to be oscillatory soon after activation. In the presence of feedback, this effect can be produced if either the natural period of the seismometer is lengthened or the feedback filter corner period is shortened beyond design values. It is probable that the natural period of the seismometer was lengthened from 15 seconds to approximately 60 seconds as a result of vibration effects. Acceptable operation has been achieved by removing, through ground commands, the feedback filters from all three components. In this configuration, the seismometers have responses equal to underdamped pendulums with natural periods of 2.2 seconds .

Figure 3-5 Passive seismic experiment and the experiment central station in the foreground with the undeployed thermal ion detector experiment in the background.

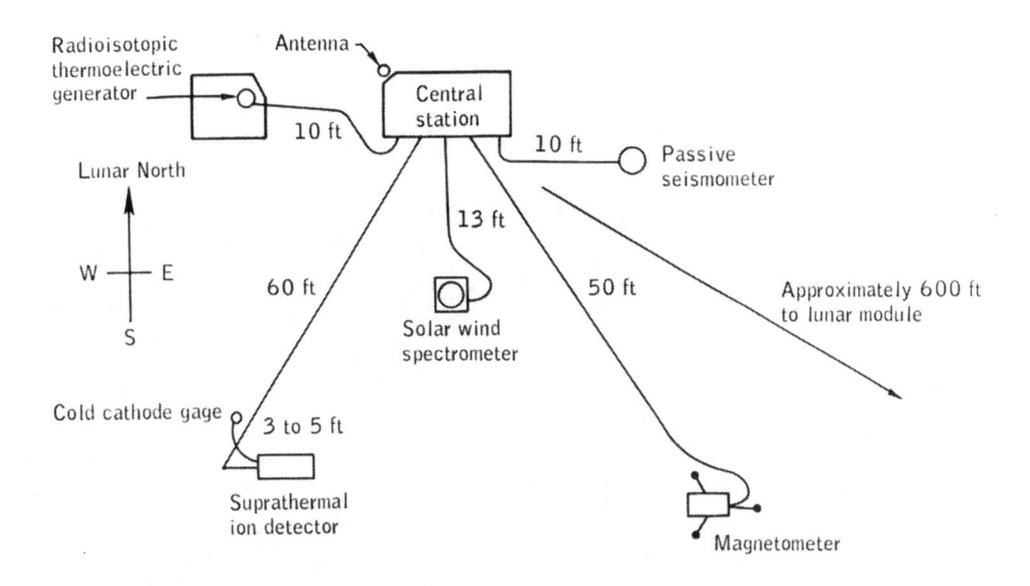


Figure 3-6 Deployment configuration of the Apollo lunar experiments package.

The active thermal control system was designed to maintain a temperature level of 125° to within 1°. The observed range is from 85° F during the lunar night to 132.5° F during the lunar day. This temperature variation will not degrade the quality of seismic data, but it will reduce the probability of obtaining useful long-period (tidal) data.

Recorded seismic signals.- Prior to lunar module ascent, a great many signals were recorded and corresponded to various crew activities, on the surface and within the lunar module. The crewmen's footfalls were detectable at all points along their traverse, with a maximum range of approximately 360 meters. Signals of particular interest were generated by static firings of the reaction control thrusters and the ignition of the ascent engine, as shown in figure 3-7. These signals traveled from their sources to the seismic sensors with a velocity of approximately 108 meters/sec. Spectra of the thruster signals show peak signal amplitudes near 8 hertz, as was observed during Apollo 11 static firings.

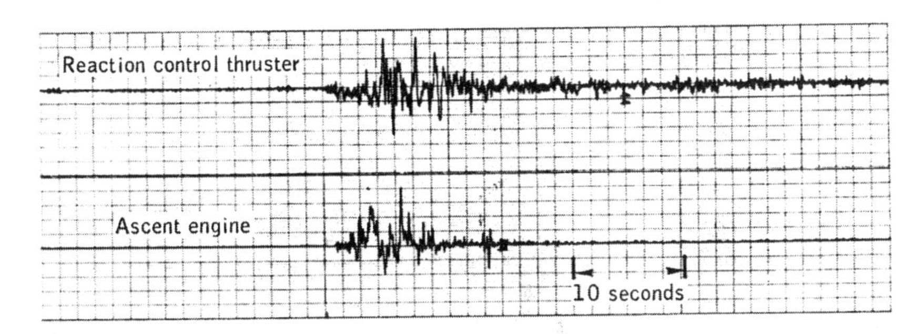


Figure 3-7 Seismic signals during reaction control thruster and ascent engine firings.

Following ascent, 18 seismic signals that could possibly be of natural origin have been identified on the records for the 10-day period of observation. All but one of the 10 high-frequency events detected by the short-period vertical component were recorded within 8 hours after lift-off and probably correspond to venting processes of the lunar module descent stage. These data contrast sharply with the hundreds of signals assumed to be of lunar module origin recorded during the first 8 days of Apollo 11 seismometer operation. This drastic reduction in the number of interfering noises from the lunar module is attributed primarily to the increase from 16.8 meters to

130 meters in distance from the descent stage. However, the reduced sensitivity of the vertical component in the short-period seismometer is certainly a contributing factor.

Of the eight signals recorded on the long-period components, three are extremely small, possibly of instrumental origin, and the remaining five are quite definite. All signals exhibit emergent onset rates and durations lasting from 10 to 30 minutes; periods which are long compared to similar seismic events on earth.

The most significant event recorded was the impact of the lunar module ascent stage at a distance of 75.9 kilometers and an azimuth of 114 degrees east of north from the experiment. The angle between the impact trajectory and the mean lunar surface was 3.7 degrees at the point of impact, and the approach azimuth was 306 degrees. Signals from the impact were recorded well on all three long-period seismometers. The signal amplitude built up gradually to a maximum of 10 millimicrons peak-to-peak on all components over a period of about 7 minutes and thereafter decreased very gradually into the background, the total duration being about 50 minutes. Distinct phases within the wave train are not apparent. The signal is shown on a compressed time scale in figure 3-8, and no phase coherence between components is evident. The spectral distribution of the signal ranges from approximately 0.5 hertz to the high frequency limit of 2 hertz for the long-period seismometer.

The seismic wave velocity, corresponding to the first arrival, ranges between 3.0 and 3.78 km/sec. The unexpectedly long duration of the wave train is assumed to have either resulted from a prolonged effective source mechanism or from a propagation effect. An extended source from such an impact might result from: (1) triggering of rock slides within a crater located near the point of impact; (2) the distribution of secondary impacts which would presumably rain downrange, and toward the seismic sensors, from the primary impact point; and (3) the effects of an expanding gas cloud consisting of residual ascent stage fuel and volatilized ejecta. If the signal duration is a propagation effect, the quality factor (Q), the lunar material through which these waves propagate must range between 2000 and 4500, as opposed to Q,-values of between 10 and 300 for most crustal materials on earth. Further interpretation of this very unusual signal must be deferred pending a final analysis. It should be noted, however, that the impact signal is similar in character to a number of prolonged signals detected by the Apollo 11 seismometers. This similarity eliminates an earlier suspicion that the Apollo 11 signals might be of artificial origin.

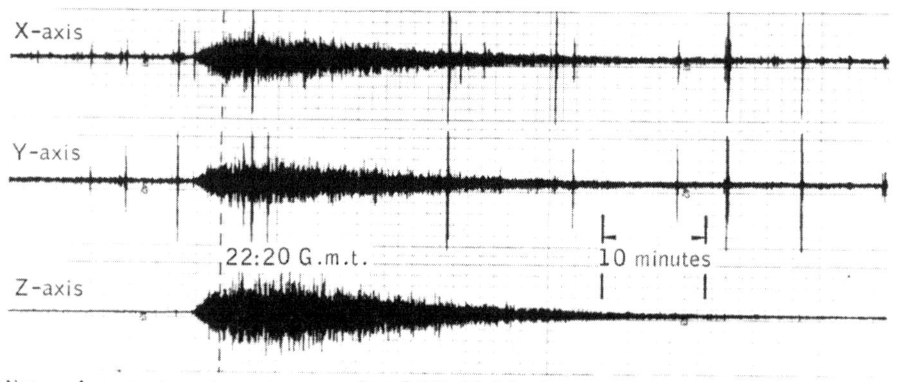


Figure 3-8 Long-period seismometer response to ascent stage impact.

A direct correlation has been made between signals recorded by the magnetometer and those recorded by the short-period vertical component. This correlation was particularly noticeable during passage of the moon through the transition zone between the tail of the earth's magnetic field and interplanetary space, where rapid variations in the magnetic field strength are observable from the magnetometer record.

Feedback outputs.- The long-period seismometers are sensitive to both tilt horizontal components and changes in gravity (vertical component). These data are transmitted on separate data channels, referred to as "feedback," or "tidal," outputs. A particularly interesting case of tilting has been observed, beginning approximately 8 hours before terminator crossing and lasting 24 hours thereafter, as shown in figure 3-9. A total tilting of 45 seconds of arc, downward and in the direction of east-north-east, occurred during this interval. The tilting may have been produced by a combination of thermal effects either on the very near lunar surface or on the instrument itself, and possibly by the tilting of large blocks of the igneous rock underlying the regolith, which is estimated to range between 1 and 5 meters in thickness. Thermal effects could not have propagated for more than a few inches into the regolith during the period of observation. Thus, tilting of underlying blocks by thermal effects would have to be produced by changes in temperature at exposed crater walls. The crew reported seeing zones of lineations 5 to 30 meters wide trending approximately north-south in this region. Such zones may have been produced by sifting of regolith material into underlying fractures.

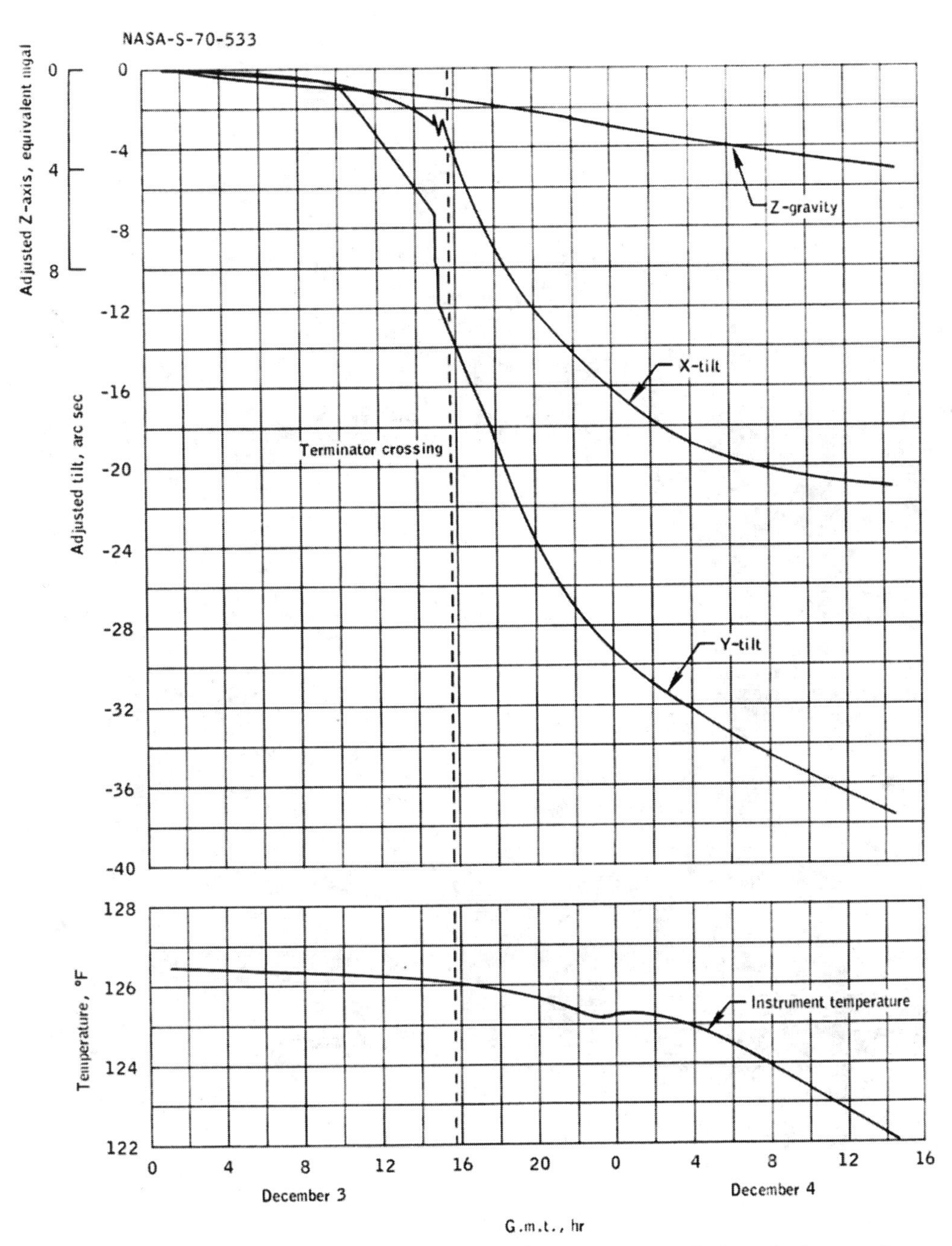


Figure 3-9 Seismometer feedback response and temperature variations during terminator passage at the landing site.

### 3.1.3 Magnetometer Experiment

she magnetometer experiment measures the magnetic field on the lunar surface in response to the moon's natural electromagnetic fields in the solar wind and the earth's magnetic tail. Measurement of the field vector and gradient permits placement of an upper limit on the permanent magnetic moment of the moon and also allows inhomogeneities and local field sources to be studied. Vector field measurements taken during the moon's passage through the neutral sheet in the geomagnetic tail will also allow determination of the moon's bulk magnetic permeability. Simultaneous field measurements taken by the lunar surface magnetometer and a lunar orbiting satellite will be used to differentiate the sources producing the lunar induction magnetic field and to calculate the bulk electrical conductivity.

The initial data show that a portion of the moon near the Apollo 12 landing site is magnetized. The data also show that the magnetic field on the lunar surface has frequency and amplitude characteristics which vary with lunar day and night. These two observations indicate that the material near the landing site is chemically or electrically differentiated from the whole moon.

Figure 3-10 Lunar surface magnetometer deployed.

The magnetometer was deployed in approximately 3 minutes, and figure 3-10 shows the deployed magnetometer at the experiments package site. Magnetic-field data were received immediately after instrument activation, and ground commands were sent to establish the proper range, field offset, and operational mode for the instrument. The experiment was deployed so that each sensor is directed about 35 degrees above the horizontal. The Z sensor is pointed toward the east, the X sensor toward the northwest, and the Y sensor completes a right-hand orthogonal system. Instrument measurements include both time-invariant and time-varying vector field information. The time-invariant fields are produced by a source either associated with the entire moon or in combination with a possible localized source. The time-varying vector fields are produced by the sun's magnetic field in the solar wind and by the earth's magnetic field in the regions of the magnetic bow shock, transition zone, and the geomagnetic tail. These regions and the moon's first orbital revolution after deployment are shown in figure 3-11. At the time of instrument activation, the moon was just inside the earth's magnetic bow shock.

The magnetic field measured on the lunar surface is a vector sum of the fields from the lunar, terrestrial, and solar magnetic fields. The selenomagnetic field associated with a local portion of the moon should have small-amplitude variations over time periods on the order of days and can therefore be separated from the higher frequency transients by measurements taken daring a period of one complete revolution around the earth. A preliminary analysis of a field measured during half an orbital period shows that the field is approximately 30 gammas in magnitude and is directed downward approximately 50 degrees from the vertical toward the southeast. The magnetic-field gradient was measured to be less than $10^{-3}$ gammas/cm in the plane tangent to the lunar surface. Magnetic-field measurements from the lunar orbiting Explorer 35 spacecraft indicate that the dipole moment is less than $10^{20}$ gauss-$cm^3$, which implies the 30-gamma field is caused by a localized source near the Apollo 12 landing site, rather than from a uniform dipole moment associated with the whole moon.

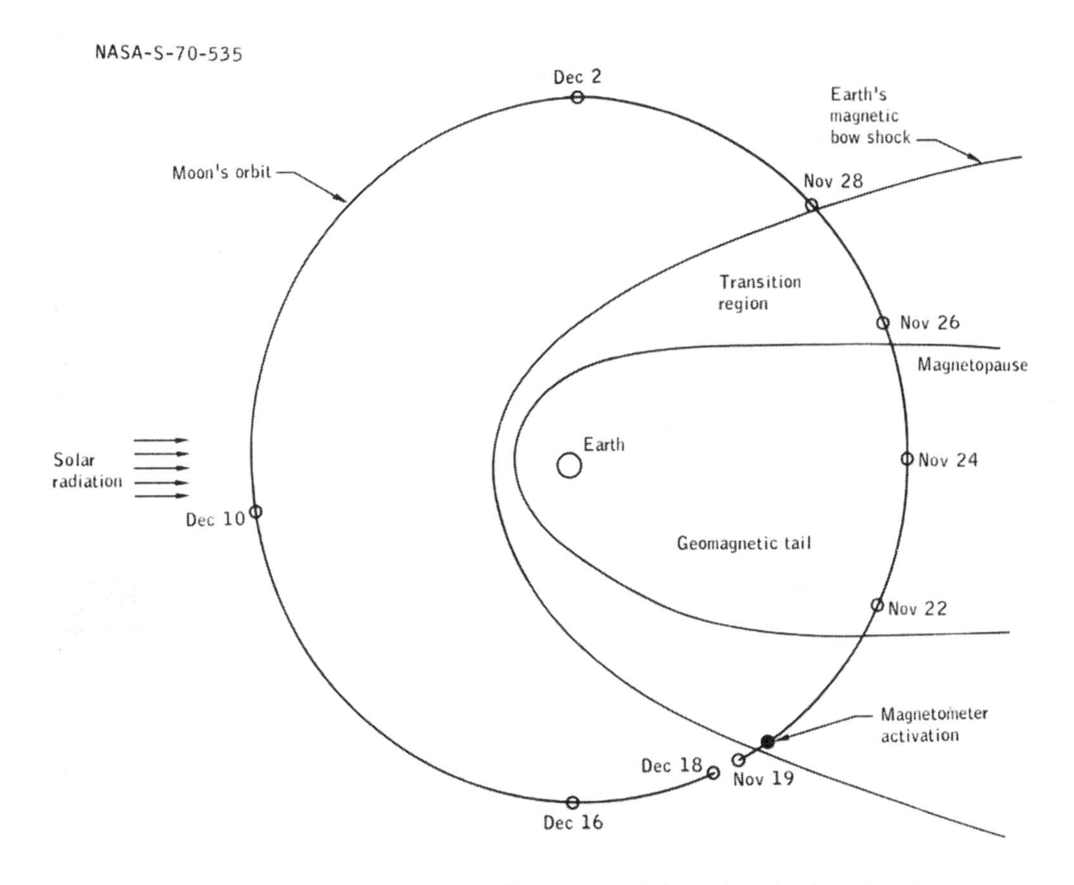


Figure 3-11 Geometry of the earth's magnetic field regions in the solar plasma.

Along with the time-invariant magnetic field associated with the moon, a relatively large time-varying component exists. During each orbit around the earth, the moon is embedded in each of the different magnetic-field regions shown in figure 3-11. The magnetic-field environment is dominated by the solar wind in interplanetary space, by the interaction of the solar wind and the earth's magnetic field in the bow shock and transition region, and by the earth's intrinsic field in the geomagnetic tail region.

Figures 3-12 through 3-15 show typical field measurements obtained during a 6-minute period in each of the three regions shown in figure 3-11. Figure 3-12 is a time-series plot of the three vector components of the magnetic field in the instrument coordinate system while the moon was in interplanetary space and the instrument was in sunlight. The field variations are caused by the fluctuating solar field transported to the lunar surface by solar

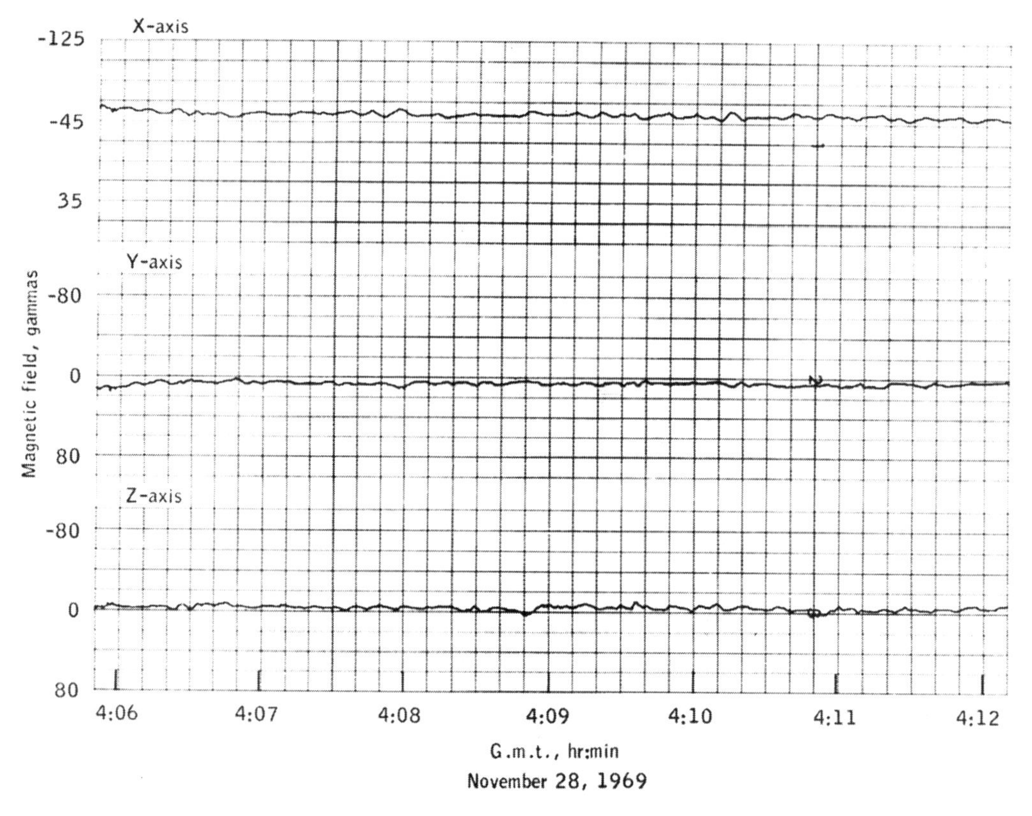


Figure 3-12 Interplanetary field region on the lunar surface in sunlight.

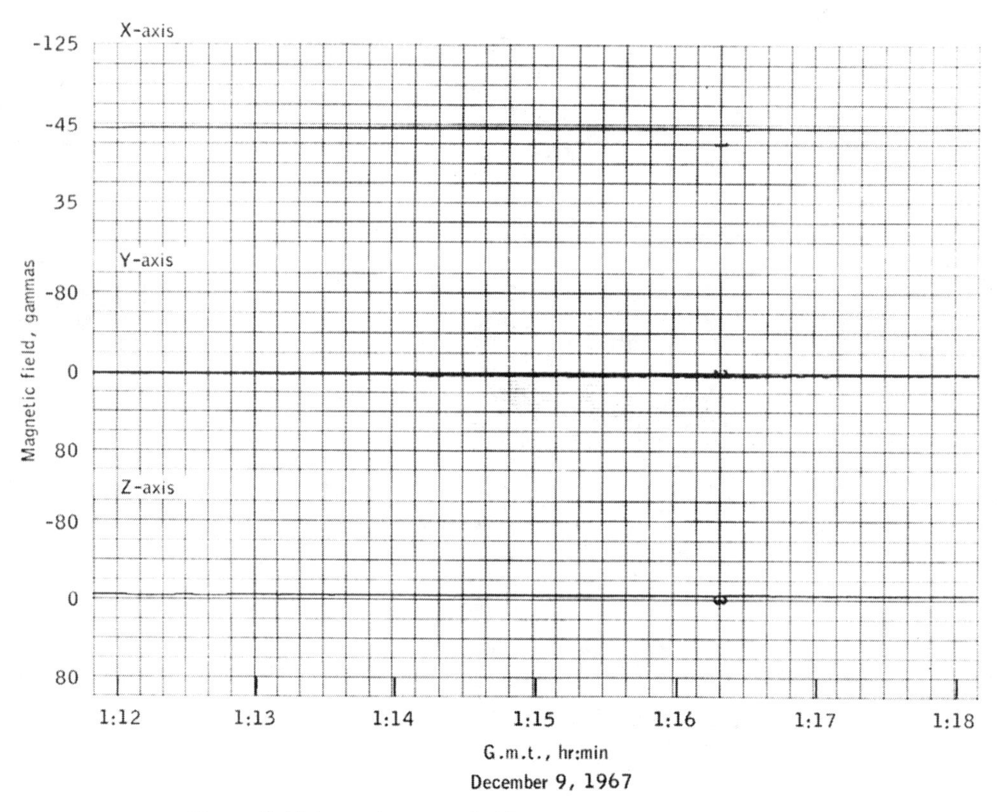


Figure 3-13 Interplanetary field region on the lunar surface in darkness.

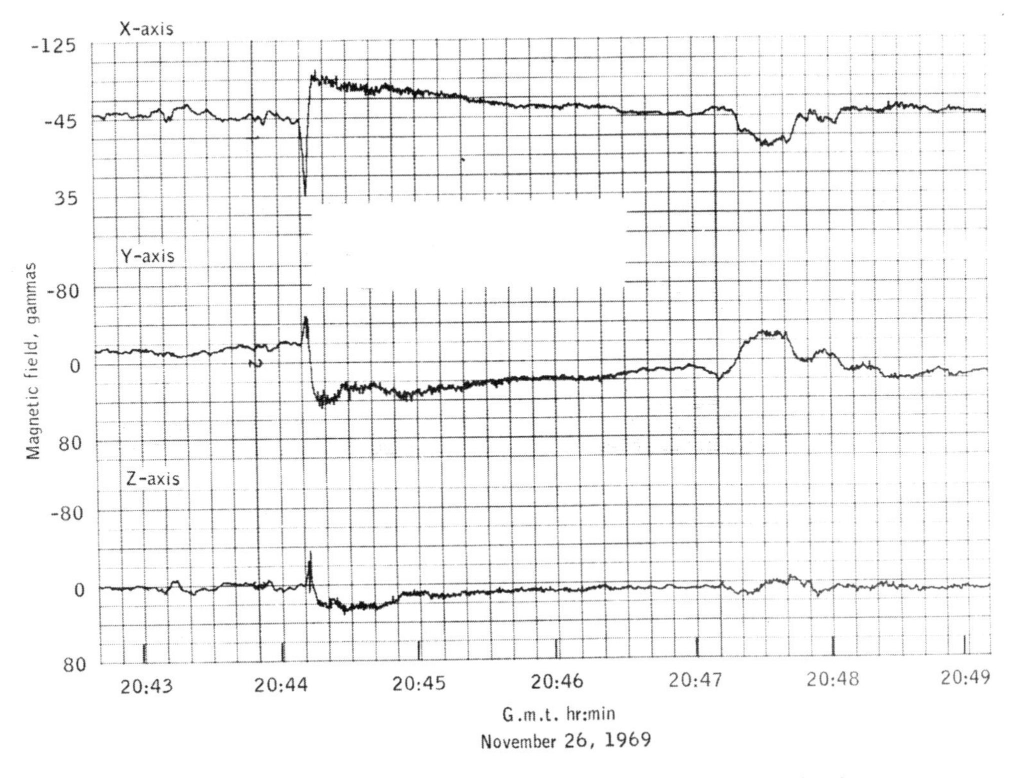


Figure 3-14 Instrument crossing of the earth's bow shock.

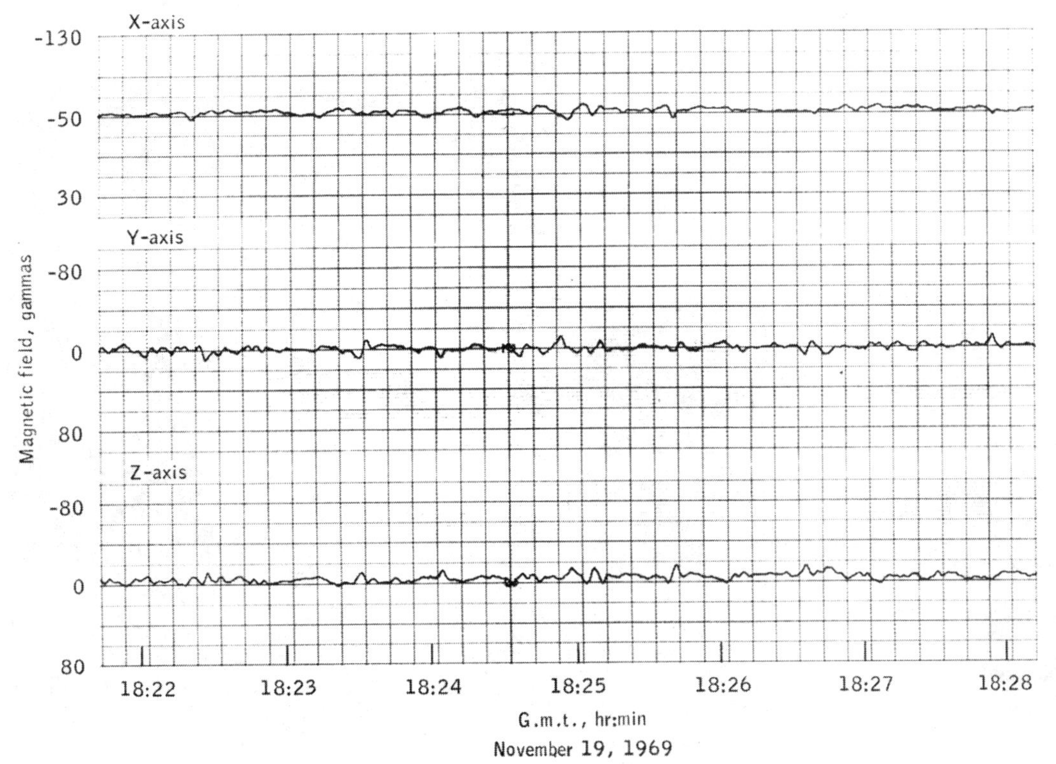


Figure 3-15 Instrument passage through the transition region between the magnetopause and the earth's bow shock.

plasma and correlate in time with data from the solar wind spectrometer (section 3.1.4). Figure 3-13 is a plot of the three vector components during a period when the moon was in interplanetary space and the magnetometer was in darkness. The resultant lunar surface field can be seen to lack the short-period fluctuations appearing in data received when the instrument was in sunlight. The magnetic-field vector components during a time when the moon was in the vicinity of the earth's plasma magnetohydrodynamic bow shock are shown in figure 3-14. The response amplitude in this region is large. Typical measurements obtained in the transition region between the bow shock and the magnetopause are plotted in figure 3-15. In this region, the field fluctuations are of greater amplitude and contain higher frequencies than in the interplanetary solar field regions. These measurements also correlate well with data from the solar wind spectrometer. As expected, measurements taken in the field region of the geomagnetic tail show very low amplitude and frequency fluctuations with time.

Temperatures measured at five different locations in the instrument were approximately 68° F higher than expected because of lunar dust on the thermal control surfaces.

Two anomalies have been observed in the operation of the magnetometer since deployment. Following discovery of a malfunction, one of the three digital filters in the data processing electronics was bypassed by ground command 3 days after equipment activation. The problem was discovered as a faulty subroutine in the digital filter that was erroneously multiplying the data by zero. After the electronics temperature decreased from a high of 161° F to below 122° F during the lunar day, the filter was commanded back into the data link and instrument operation was satisfactory. Preliminary indications are that a welded connection parted at the upper temperature. The second anomaly occurred about 3 weeks after deployment, when the three vector-component measurements dropped off-scale and the vector magnetic field could not be measured. Subsequent commands permitted the X-component measurements to be brought back on scale but not the Y- and Z-sensor outputs. All subsystems were operating normally except for the sensor electronics. Another attempt will be made to restore the sensor electronics to proper operation when the temperature of the electronics rises at lunar sunrise.

### 3.1.4 Solar Wind Spectrometer

Since the solar wind spectrometer was activated on the lunar surface, the performance and the data received have been satisfactory. The solar wind spectrometer was turned on by ground command at approximately 122-1/2 hours. All background plasma and calibration data appear normal. The seven dust covers were successfully deployed at 143-1/2 hours.

The observed plasma ion data, characteristic of the earth's "transition region," were found to be consistent with that indicated by the magnetometer. As expected, the plasma properties are highly variable in the transition region. The bulk velocity was near 300 km/sec, the density was about 5 ions/$cm^3$, and fluxes of from 0.5 x $10^8$ to about 2 x $10^8$ ions/ $cm^2$-sec were observed. High-energy electrons were also detected.

When the instrument entered the geomagnetic tail of the earth, essentially no solar plasma was detected. Upon emerging from the geomagnetic tail, the spectrometer again passed through the transition region.

Nine days after deployment, the instrument passed through the plasma bow shock of the earth into the interplanetary solar wind, which exhibited the following typical plasma properties: bulk velocity of from 500 to 550 km/sec, density of from 2 to 2.5 ions/$cm^3$, and a flux of approximately 1.4 x $10^8$ ions/$cm^2$-sec.

With the onset of lunar night, the plasma activity, as predicted, decreases to below the measurement threshold of the instrument.

### 3.1.5 Suprathermal Ion Detector

The suprathermal ion detector experiment functioned normally until 14-1/2 hours after activation, at which time the 4.5-kV and 3.5-kV power supplies and the voltage sequencer for the low-energy curved-plate analyzer shut down. At the same time, the sequencer for the high-energy curved-plate analyzer skipped forward five data frames and returned to normal sequencing on the next cycle. After successfully commanding on the sequencer and the 3.5-kV power supply, all attempts were unsuccessful in restoring the 4.5-kV power supply.

Instrument operation continued until about 29 hours after activation, when the instrument changed its data accumulation mode, and the high-energy and low-energy sequencer voltages went to zero. The instrument was immediately commanded into the normal operating mode and the sequencers commanded back on. At this time, the total ion-detector background counts were close to 200 counts per accumulation interval and were increasing, indicating a pressure rise with temperature. For this reason an arc in the 3.5-kV power circuit to the detector was suspected and the 3.5-kV power supply was commanded off. Following lunar noon (13 days after activation) the 3.5-kV power supply was reenergized and the experiment has remained fully functional. However, daily attempts to command on the 4.5-kV power supply have been unsuccessful.

The following observations of scientific interest have been detected during the first 18 days of full operation:

a. The ascent-engine firing

b. Ascent stage impact

c. Presence of sporadic low-energy ion clouds during first passage through the earth's transition region. One typical event in this region showed the passage of an ion cloud, the beginning of which was indicated by both the detection of 750-eV ions and an associated magnetic field that was sensed by the magnetometer, with the remaining ions of the cloud generally in the energy range of from 30 to 100 eV

d. Presence of low-energy ions with narrow energy spectra, indicating the ground screen has some influence on incoming thermal ions

e. Presence of very energetic protons and/or alpha particles on the night side (fig. 3-16)

f. Presence of solar wind ions on the night side

g. A possible sunrise-related pressure wave characteristic of the moon.

h. Possible gaseous emission from the descent stage following sunrise.

The data are too preliminary to justify a detailed discussion, and a more rigorous analysis of these observations will be presented in a later science report.

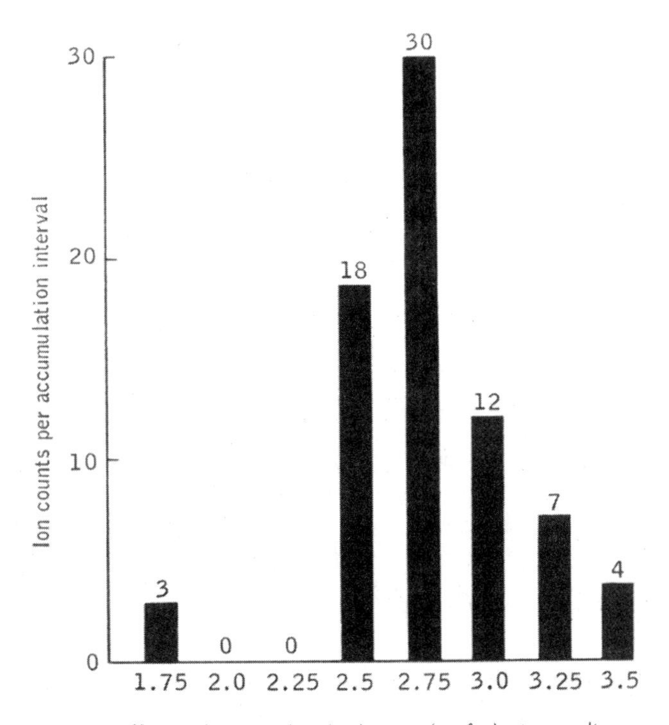


Figure 3-16 Typical high energy spectrum at 1919 G.m.t. on December 4, 1969.

### 3.1.6 Cold Cathode Gage

As expected, the cold cathode gage indicated full-scale response at activation because of gases trapped within. the instrument. After about a half hour of operation, the response changed perceptibly from the full-scale reading. After 7 hours, the indication had decreased to about 3 x $10^{-9}$ torr. At the time of lunar module derressurization prior to the second extravehicular activity period, the response increased to at least 7 x $10^{-8}$ torr. The exact value is uncertain because a programmed calibration, which time shares the data channel, was being performed near the time of maximum pressure. The pressure increase resulting from lunar module outgassing is in reasonable agreement with predictions. Whenever a crewman approached the experiment during the second extravehicular activity period, the instrument response went off-scale, as expected, because of gases released from a portable life support system.

The stiffness of the electrical cable joining the cold cathode gage to the suprathermal ion detector experiment caused some difficulty during deployment of the gage (see section 14.3.5). To avoid this problem the tape wrap will be eliminated from future experiment packages and will decrease the cable stiffness The instrument apparently suffered a catastrophic failure after about 14 hours of operation, because of a malfunction either in the 4.5-kV power supply or in the power-supply switching mechanism.

## 3.2 SOLAR WIND COMPOSITION EXPERIMENT

The solar wind composition experiment was designed to measure the abundance and the isotopic composition of the noble gases in the solar wind. In addition, the experiment permits a search for the isotopes tritium ($H^3$) and radioactive cobalt ($Co^{56}$). The experiment hardware was the same as that flown in Apollo 11 and consists of a specially prepared aluminum foil with an effective area of 0.4 square meter. Solar wind particles arrive at velocities of a few hundred kilometers per second and, when exposed to the lunar surface environment, penetrate the foil to a depth of several millionths of a centimeter, becoming firmly trapped. Particle measurements are accomplished by heating portions of the returned foil in an ultra-high vacuum system. The emitted noble gas atoms can be separated and analyzed in statically operated mass spectrometers, and the absolute and isotopic quantities of the particles can then be determined.

The experiment was deployed on the lunar surface and was exposed to the solar wind for 18 hours 42 minutes, as compared to 77 minutes for Apollo 11. Afterward, the foil was placed in a special Teflon bag and returned to earth for analysis.

## 3.3 LUNAR GEOLOGY

Geological information, in the form of voice descriptions, lunar surface samples, and surface photographs, was also provided during all other phases of the surface stay. It appears that the locations and orientations of a significant number of the returned samples can be determined relative to their positions on the lunar surface; therefore, detailed geologic maps and interpretations can be made from this information. A summary of the returned lunar surface samples, compared with the Apollo 11 samples, is contained in the following table:

| Material | Approximate weight, lb | |
|---|---|---|
| | Apollo 12 | Apollo 11 |
| Fines* and chips | 12.8 | 24.2 |
| Rocks | 61.0 | 24.3 |
| Core-tube specimens | 0.9 | 0.3 |
| Total | 74.7 | 48.8 |

*NOTE: Terms used in this section are defined in a glossary, Appendix F

### 3.3.1 Geology of the Landing Site

The lunar module landed on the southeastern part of the Ocean of Storms at 110-1/2 hours. The coordinates of the landing site are given in section 4.3. This portion of the Ocean of Storms mare is dimpled by many small craters of Copernican and Eratosthenian age, and the landing site is contained within a broad Copernicus ray. The site is located on the northeast rim. of the 150-meter-diameter Head crater and the northwest rim of Surveyor crater, in which the Surveyor III unmanned spacecraft landed on April 20, 1967. See figure 3-17 for a traverse map of the landing-site area. The surface northwest of the landing site is littered with debris from a 450-meter crater, informally called the Middle Crescent crater, the southeast rim of which lies about 200 meters northwest of the landing site.

On the second extravehicular excursion, the crew visited four craters of over 50 meters in diameter, and many of smaller size. The characteristics of eight craters were described, and a variety of material ejected from each was collected. The crew made numerous comments about smaller craters and about the surface features between them, including ground that may be underlain by ray material from more distant craters, especially Copernicus. The rock collections returned to earth contain a variety of material ejected from local craters visited on the traverses. These collections included fine-grained materials of both local origin and from far-distant sources.

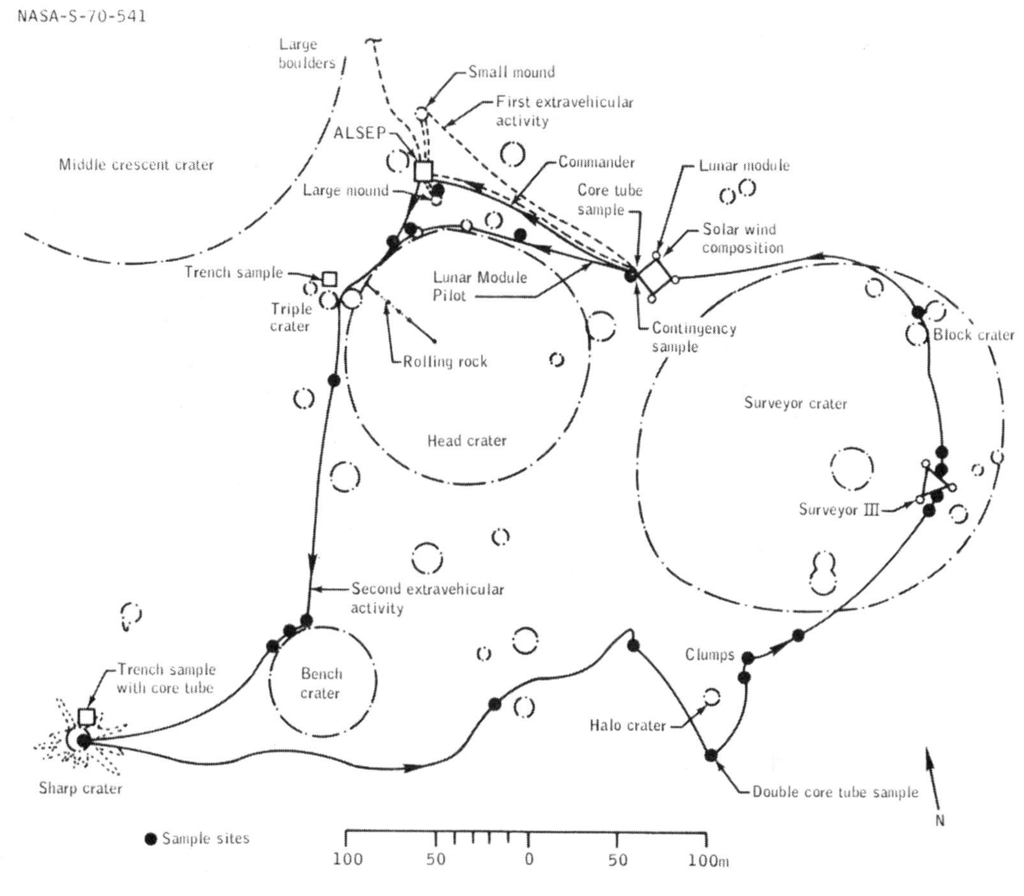


Figure 3-17 Traverse map.

Regolith.- During the landing operations, the regolith, or finegrained layered material on the lunar surface was only penetrated to an average depth of about 5 centimeters by the lunar module foot-pads. The loose regolith material beneath a crewman's boots compacted into a smooth surface. Many crew comments concerned the large amounts of glass contained in this regolith. Beads and small irregularly shaped fragments of glass were abundant both on the surface of and within the regolith. Glass is also splattered upon some of the blocks of rock at the surface and is concentrated within many shallow craters. The crew commented "Every crater you .... look in, you see glass beads."

Along many parts of the geology traverse, the crew found a finegrained material of relatively high albedo. At some places, this material is at the surface (for example, near the rim of Sharp crater) but at other localities is buried beneath 10 centimeters, or more, of darker material (as on the west side of Head crater and on the outer slope of Bench crater This fine-grained material may constitute the deposit which is observed in the telescope as one of the bright rays of Copernicus.

The darker regolith above the light-gray material is only a few centimeters thick in some places but probably thickens greatly on the rims of some craters. The darker regolith appears to show more variation from one locality to another than does the light-gray regolith. These regolith variations include differences in both the size and shape of the particles and in the observed mechanical properties. Most of these differences probably result from the effects of local cratering events. The differences in abundance, size, and angularity of ejected blocks, as well as the petrologic differences of the rock fragments on and in the surface regolith, appear to be closely related to local craters from which some of the blocks have apparently been derived.

Patterned ground was noted northwest of the lunar module, at and near Surveyor III, on the outer slopes of Sharp crater, and near Halo crater. Northwest of the lunar module, this patterned ground was described as consisting of linear traces or grooves only about 1.3-centimeter deep and probably of the same type shown in Apollo 11 photographs. The grooves are oriented north-south. These features were also observed near Middle Crescent crater at a distance of about 200 meters from the lunar module. Near Surveyor III, however, the lineations were described as having a generally northwest orientation. This phenomenon correlates with the patterned ground shown in certain lunar orbiter photographs, but the associated grooves are obviously much larger than those described in Apollo 12.

Figure 3-18 Blocky ejecta near a small crater photographed during the first extravehicular activity period.

A tentative interpretation, of the upper two layers of the regolith is suggested. The light-gray material which underlies the darker material quite possibly is ray material related to Copernicus, and the darker regolith consists partly of debris ejected from local craters younger than Copernicus. Probably there has been considerable mixing together of material from these two sources as a result of subsequent smaller cratering events. Other processes,

Figure 3-19 Photograph of Bench crater showing probable bedrock.

such as downslope creep, may also have contributed to this mixing, and later "space weathering" processes may have contributed to the change in surface albedo.

Craters and block fields.- The supposition that the darker regolith is largely of local origin is strengthened by crew observations of the larger local craters and their block fields. Information on the distribution, size, shape, abundance, and petrologic dissimilarity of the blocks observed in different areas of the traverse is particularly pertinent in an interpretation of the remainder of the regolith.

Northwest of the lunar module is Middle Crescent crater, the largest visited. The crew observed huge blocks on its wall, probably derived from the local bedrock. According to one crewman, blocks on the surface between this crater's rim and the lunar module consist of "everything from fine-grained basalts to a few coarse-grained ones."

Both rounded and angular blocks were found on the western edge of Head crater and described. One rock the size of a grapefruit was tossed into the crater to excite the seismometer and went skipping and rolling down the slope in slow motion. Most rock fragments were angular and of a dark gray color (fig. 3-18). These blocks were reported to be much more abundant on the rim nearest the crew than on other parts of the rim. Some rocks appeared to be coarse in grain and their crystals showed clearly, even when covered with lunar surface material. These crystals were described in one of the rocks as being a very bright green, much like a "ginger ale" bottle. The crystals are obviously basalts and coarsergrained rocks that were ejected from Head and Middle Crescent craters.

Bench crater appears to show some significant differences in its ejecta and morphology. Numerous large blocks were apparently ejected from this crater, some as large as a meter in length. These rocks, some angular and others rounded, were estimated to make up 5 percent of the material surrounding the crater. Material in the bottom of the crater was reported most likely to be bedrock (fig. 3-19) and appeared to have been molten at one time. Numerous "glass beads," some of which were collected, were reported to be on the sides and in the vicinity of this crater. The crater derives its informal name from a bench-like protrusion located high on the crater wall and apparently totally free of regolith. This protrusion remains unexamined because the steep slope of the crater walls prevented a closer investigation.

Blocks observed on the south rim of Surveyor crater and near Surveyor III are quite similar to those from Head and Middle Crescent craters. Angular blocks, some cube- and others brick-shaped, were also noted near Surveyor III. One rock was described as having shear faces and abrasion marks on it, and it also contained the bright crystals.

Photographic panoramas were taken across the 10-meter-diameter crater (informally called "Block" crater) within Surveyor crater. Nearly all the blocks from this crater were described as sharply angular. The sharp angularity of the blocks suggests that the crater is relatively young.

Sharp crater contrasts strikingly with the blocky-rim craters previously described. It is a small crater with a rim, less than a meter high, composed of high-albedo material, which has also splashed out radially. The core tube driven in the rim of the crater penetrated this ejecta without difficulty.

The Halo crater area seems to contain a group of small craters that are without block fields. Little description of this area was reported, aside from the fact that a patterned ground, with a coarse texture of ripples and dimples, was present.

Figure 3-20 Mound just north of Head crater as viewed from the northeast.

The crew reported observing two unusual mounds just north of Head crater. The larger of these mounds was scoop-sampled and was later determined from photographs to be about 1.3 meters high, 1.5 meters in diameter at the top, and about 5 meters in diameter at its base (fig. 3-20). These mounds (fig. 3-21) are probably composed of slightly hardened clods of fine-grain-:. material that was ejected from one of the nearby craters.

Figure 3-21 Material on top of a reported mound.

### 3.3.2 Mechanical Properties

Crew observations, photography, telemetered dynamic data, and examination of the returned surface samples permit a preliminary assessment of the physical and mechanical properties of these materials and a comparison with Apollo 11 results.

Descent and touchdown.- Lunar surface erosion resulted from descent-engine exhaust gases, and dust was blown from the surface along the trace of the final descent path (see section 6). Examination of sequence-camera film suggests that this erosion was greater than observed in Apollo 11. Further analysis is required to ascertain whether this effect resulted from different surface conditions, a different descent profile, or whether degraded visibility resulted from a different sun angle.

The landing was gentle, causing only limited stroking of the shock absorbers. The plus-Y footpad apparently contacted the surface first (see section 4.2) and bounced a distance of about one pad-width. The minus-Y footpad slid laterally about 15 centimeters and penetrated the soil to a depth of about 10 or 12 centimeters. The other footpads penetrated to depths of from 2 to 5 centimeters, as typically shown in figure 3-22. Similar penetrations were observed under similar landing conditions at the Apollo 11 site, indicating that the surface material bearing capacities at the two sites are of the same order of magnitude.

Extravehicular activity.- After an initial acclimation period, the crew encountered no unexpected problems in moving about on the surface. Traction appeared good, and no tendency for slipping or sliding was reported. Fine surface material was kicked up readily and, together with the lunar dust that coated most contacting objects, created difficult working conditions and housekeeping problems on board the spacecraft (section 6).

Footprint depths were of the same order as in Apollo 11, that is, a centimeter or less in the immediate vicinity of the lunar module and in the harder lunar surface material areas, and up to several centimeters in the softer lunar surface material areas. The least penetration was observed on the sides of Surveyor crater. Penetration of the lunar surface by various handtools and staffs was reported as relatively easy and was apparently easier than reported for Apollo 11. The staff of the solar wind composition experiment was readily pushed to a depth of approximately 11 centimeters and the flagpole approximately 17 centimeters. Trenches were dug to depths of 20 centimeters without difficulty, and the crew reported that, except for limitations caused by the lengths of the tool handles (section 9), they could have excavated to considerably greater depths without difficulty. Vertical sidewalls on these trenches would cave in when disturbed at the top but would remain vertical if left untouched.

Figure 3-22 Detail of lunar module minus Z footpad showing disturbance of fine-grained material as viewed from the east.

Core tubes were pushed and driven at three sites (see fig. 3-17); single core-tube specimens were taken near the lunar module and in the bottom of a trench at Sharp crater, and a double core-tube specimen was retrieved at Halo crater. In both of the single-tube specimens, the tube was easily driven to its full depth. The double core-tube specimen was taken to a depth of approximately 70 centimeters. The core tubes were easily withdrawn, and the holes remained open unless disturbed. The interior design of the core-tube bits was different from that of Apollo 11, in that the Apollo 12 internal diameter was constant. This redesign probably contributed to the ease with which they were driven.

No change in the texture or consistency of the lunar material with depth was observed during trenching or the driving of core tubes. As expected, the subsurface material is darker than the suface material, except in the area just northwest of Head crater where subsurface material was lighter.

The following conclusions regarding three distinct areas, in terms of lunar material texture and behavior (fig. 3-17), were made by the crew: (1) the region between Halo and Surveyor craters, including the inside slope of Surveyor crater, has the firmest surface material and the appearance of ground upon which light rain has fallen; (2) the vicinity of Sharp crater has the softest surface material and permits the deepest footprints; and (3) the vicinity of the lunar module has lunar material intermediate in character. The probing of portions of the protruding features described as "mounds" revealed a composition of fine-grained compacted material which crumbled easily.

Examination of the photographs taken at the Surveyor III site (figs. 3-23 and 3-24) suggest that the lunar surface has undergone little change in the past 2-1/2 years. The trenches excavated by the lunar material sampling device on Surveyor, as well as the waffle pattern of the Surveyor footpad imprint, appear much the same as when formed on Surveyor landing (fig. 3-25). Many of the Surveyor components (fig. 3-26) were observed to be coated with a thin layer of dust, but some other process could also have discolored them. The results of a detailed postflight examination of the Surveyor components returned to earth will be published in a separate science report (see appendix E). The Surveyor components returned were a cable, a painted tube, an unpainted tube, the television camera, and the scoop.

Examination of returned samples.- Four kilograms of lunar surface material having a grain size of less than 2 millimeters in length was returned and this was much less than the 11 kilograms returned from Apollo 11. The lunar surface samples available for study are: (1) lunar surface material mixed with and adhered to the rock samples in both the selected and documented sample boxes; (2) five individual documented lunar material samples; (3) the contingency sample; and (4) the contents of four core-tube specimens. A cursory examination of returned samples indicates a very fine, dusty, charcoal-gray lunar material similar to that returned from Apollo 11.

Only one of the documented lunar surface material bags has been opened. This sample was taken in a trench dug in the northwest quadrant of Head crater and has a distinctly different color from the other lunar material samples in that it is light gray, similar to the color of cement. The lunar material in the contingency sample bag weighs approximately 1100 grams but has not yet been examined.

Thus far, only one core-tube sample, that taken during the first extravehicular excursion in the vicinity of the lunar module, has been opened and examined. This core sample was 19.4" centimeters long, and its average bulk density was calculated to be 1.73 grams/cm$^3$. The Teflon follower was found to be wedged in one-half of the inner split-tube. Because the core tube was driven into the lunar surface to its entire length of 35 centimeters, the stuck follower probably prevented a longer sample from being recovered. The medium to dark-gray color of the core sample was essentially the same as that seen in Apollo 11. The grain size distribution was also similar, with about 50 percent of the sample being finer than 0.08 millimeter.

NASA-S-70-547

Figure 3-23 Surveyor III photographed from the south.

NASA-S-70-548

Figure 3-24 Surveyor III with the lunar module in the background.

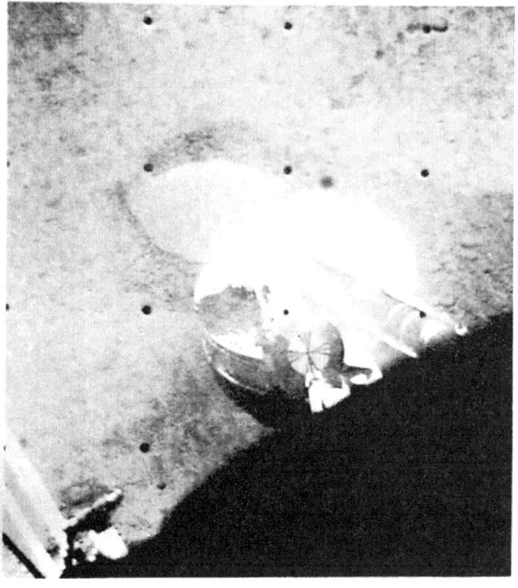

(a) Surveyor television photograph transmitted soon after landing (April 1967).

(b) Apollo 12 photograph (November 1969).

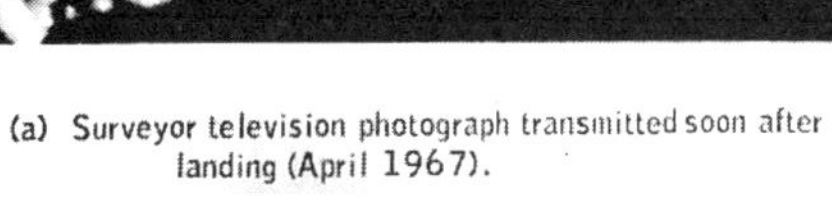

Figure 3-25 - Detail of a Surveyor III footpad showing imprints and local surface conditions

Figure 3-26 Closeup of Surveyor III.

### 3.3.3 Geologic Handtools

The handtools used during extravehicular activity were nearly identical to those for Apollo 11, and their performance is discussed in section 9. One aspect not reported by the crew was the difficulty in determining from voice communications whether the crew was reporting the Letter B or D from the sample bag numbers. For future missions, the bags will be identified so that when the number is reported by voice, it is not ambiguous when received on the ground.

## 3.4 EXAMINATION OF RETURNED SAMPLES

The bulk of the preliminary examination planned for returned lunar samples has been completed, and precautionary exposure of all the biological test systems has been conducted so that sample release can occur on schedule.

## 3.5 PHOTOGRAPHY

During the mission, all but two of the total of twenty-five 70-mm and 16-mm film magazines carried on board were returned exposed. A partially exposed 70-mm magazine had jammed and was inadvertently left on the lunar surface, and one 16-mm magazine was not used. Approximately 53 percent of the suggested targets of opportunity from lunar orbit were photographed.

### 3.5.1 Photographic Objectives

The lunar surface photographs included:

a. Long-distance photography from the command module during translunar and transearth coast for documentation purposes

b. Surface photography from lunar orbit, including multispectral strip photography and selected targets of opportunity for selenographic purposes and for use in planning and training for future missions

c. Photography of the lunar surface during descent and ascent

d. Sextant photography of the Lansberg area from orbit

e. Photography of the lunar module and experiment equipment

f. Photography of the crew performing various lunar surface tasks

g. Photography of the surface environment

h. Panoramic and stereo photographs of samples, sample areas, selenogolic features, and the traverse regions for documented scientific study

i. Photography of selected portions of the Surveyor III spacecraft and surrounding surface.

### 3.5.2 Film Description and Processing

Special care was taken in the selection, preparation, calibration, and processing of film to maximize returned information. The types of film included and exposed are listed in the following table:

| Film type | Film size, mm | Magazines | ASA speed | Resolution, lines/mm High contrast | Low contrast |
|---|---|---|---|---|---|
| SO-368, color | 16 | 12 | 64 | 80 | 35 |
| | 70 | 2 | | | |
| SO-168, color | 16 | 2 | * | 63 | 32 |
| | 70 | 2 | | | |
| SO-164, black and white | 16 | 1 | 10 | 170 | 65 |
| 3400, black and white | 70 | 4 | 40 | 170 | 70 |
| SO-267, black and white | 70 | 2 | 278 | 85 | 38 |

* Exposed and developed at ASA 1000 for interior photography and ASA 100 for lunar surface photography.

### 3.5.3 Photographic Results

Orbital photography.- For the first time during an Apollo mission, areas of the western portion of the moon's front face were in sunlight. This illumination permitted a large amount of photographic coverage which complements previous results.

Two terminator-to-terminator photographic strips were accomplished using the 70-mm still camera with an 80-mm lens. The camera was mounted on a bracket in the rendezvous window and timed by an intervalometer, which triggered exposures every 20 seconds. One strip, extending from 122 degrees east to 52 degrees west longitude along the lunar ground track, was taken on the 40th lunar orbit revolution. The second strip, taken during revolution 44, was stopped at 37 degrees east longitude because of the necessity to accomplish landmark tracking and to repeat some high-resolution photography in the next revolution. The quality of the strips, including overlap, exposure, and simultaneous 16-mm sextant photography was good and fulfills the intended mission objectives (see section 12).

Three potential landing sites, near the lunar surface areas Fra Mauro, Descartes, and Lalande, and their approach paths were photographed in stereo on one of the 80-mm strips with the 500-mm lens. The imagery is considered, at best, of fair quality. While window and lens transmission effects, as well as possible lens vibrations, affected the quality of the photography, the main cause was the high sun angle resulting from the photographs being taken on a later orbit than planned. The high sun angle created a softer image with less shadow definition, which naturally degrades the information content.

Fra Mauro was photographed with the 80-mm lens at a low sun angle, which shows the amount of shadow that can be expected during a lunar landing at this site.

The 16-mm photography taken from the command module includes good lunar surface strips taken from the window and through the sextant, tracking sequences through the sextant, and certain lunar module orbital maneuvers. Included are strips showing Lalande, Descartes, Fra Mauro, and the Apollo 12 landing area.

Surface Photography.- The lunar terrain over which the lunar module traveled during descent was documented by the 16-mm sequence camera. Lunar surface visibility during descent and the obscuration by dust just prior to landing are illustrated in this film sequence (fig. 6-1). The 70-mm film exposed on the surface, when not affected by sun glint on the lens or surface washout by sunlight, was generally of good quality.

Crew activities and lunar surface features near the lunar module, the experiment package, and those observed during the two extravehicular excursions were well documented by still-camera short sequences and by a number of panoramic views.

## 3.6 MULTISPECTRAL PHOTOGRAPHY EXPERIMENT

Inspection of the prints from the multispectral four-camera photography array indicates that the experiment was performed as planned. In addition to photography of three planned targets of opportunity using the experiment camera, continuous vertical strip photography was obtained from the command module from 118 degrees east to 14 degrees west longitude. A total of 141 pictures was taken with each of the red-, green-, and blue-filter cameras and approximately 105 with the infrared-sensitive camera. Included in the frames are a wide variety of lunar surface features, which should allow an excellent demonstration of the multispectral techniques developed in Apollo 9 (see reference 3) for lunar application. The lunar multispectral photography will provide the first high-resolution look at subtle color variations on the lunar surface, as well as the first study of color behavior at and near the zero-phase point.

An error in the preflight determination of exposure settings resulted in overexposure of approximately 30 frames in the second portion of photography conducted during the twenty-seventh lunar orbit revolution. However, almost all the data in these frames are recoverable, since maximum and minimum densities for all frames generally fall within the straight line portion of the film characteristic curve.

The assigned targets of opportunity did not fall in the center of the frame for photography of the potential landing sites Descartes and Fra Mauro. Although the targets are within the frames, the misalignment of the spacecraft was on the order of 10 or 15 degrees.

### 3.6.1 Petrology

The samples are composed primarily of igneous rocks exhibiting a wide variety of textures and compositions. The rocks range from finegrained scoria, clearly of volcanic origin, to coarse-grained pegmatitic gabbros. Differences in texture and major components suggest that the collection represents a series of cumulates in a stratified flow of basaltic composition.

Modal compositions range from anorthositic to rocks containing 30 percent olivine. Opaque content is variable but generally lower than for the Apollo 11 samples.

Ilmenite, trachyte, and free iron occur, indicating a nearly nonexistent or absent oxygen environment during crystallization. High-temperature quartz polymorphs occur in many of the igneous rocks. Sanidine has been identified in one of the breccias.

The mafic minerals, olivine and pyroxene, indicate a high-temperature environment at one time. Olivine is fayalitic, and some grains Contain 5 moles of calcium oxide, a high-temperature composition. Pigeonite is the dominant pyroxene and is iron rich, also indicating a high temperature in the parent melt.

No indication of hydrous alteration of any samples has been observed.

Samples of fines in the documented sample return container have structures suggestive of explosive volcanic origin. Several fragments appear to be pumice, and their color is generally lighter than for typical lunar soil.

### 3.6.2 Chemistry

Emission spectrographic analyses have been completed on a series of igneous rocks and several samples of fines. Silicon dioxide content averages 40 percent. Titanium dioxide content ranges from 3 to 5 percent in the igneous rocks and as high as 8 percent in the fines. Potassium oxide content is generally low, ranging from 0.04 to 0.08 percent. No potassium oxide was detected in several tested samples. These values are considerably lower than values for Apollo 11 samples.

Uranium and thorium concentrations in the igneous rocks are unusually uniform. Uranium averages 0.24 parts per million and thorium 0.9 parts per million, values which are considerably less than for Apollo 11. However, radioactive potassium, uranium, and thorium contents are significantly higher in a breccia sample than for Apollo 11.

The total carbon contents in a sample of igneous rock and part of the biocontrol sample were reported as approximately 100 parts per million (probably representing indigenous material) and approximately 600 parts per million, respectively, and these quantities represents a significant amount of carbon contamination incurred during processing.

A noble gas analysis indicates amounts of rare gases similar to the Apollo 11 results. Although argon measurements, coupled with potassium values, suggest that the Apollo 12 site is somewhat younger than the Apollo 11 site, the exposure ages ranging from 10 to 100 million years are comparable to Apollo 11.

## 4.0 LUNAR DESCENT AND LANDING

The factors influencing the selection of the Apollo 12 landing site, the actual landing operation, and the final determination of the landing site coordinates are discussed. A more detailed discussion of the landing site selection process will be published in a supplemental report (see appendix E).

### 4.1 LANDING SITE SELECTION

Two major considerations influence the selection of lunar landing sites: (1) operational and scientific objectives, and (2) launch window factors, which are related to both spacecraft performance and operational constraints. This section discusses those aspects of landing site selection significant to Apollo 11 and 12 mission planning.

### 4.1.1 Site Selection Criteria

Landing site selection for any lunar mission involves the consideration of various operational constraints, crew training requirements, terrain analyses, constraints on the preparation of support products (such as maps and models), and mission objectives. Because of the lead-time necessary to meet several of these requirements, the Apollo 12 site had to be chosen prior to the Apollo 11 launch. The site chosen had to be such that it could take advantage of an Apollo 11 success and thereby represent the next reasonable step in the lunar exploration program; at the same time provisions had to be made to land at a less ambitious site in the event Apollo 11 was not successful. The discussion of this selection process and its evolution will be presented in detail in a supplement to the mission report (appendix E).

Because of a lead time of 5 months prior to launch, the initiation time for launch-vehicle targeting corresponding to an Apollo 12 November launch occurred before Apollo 11 lift-off. After the Apollo 11 success, site selection for Apollo 12 was greatly simplified. Of the four candidates (sites 2, 3, 5, and 7), site 5 was the most desirable backup site for Apollo 12. Site 7 was selected based on satisfying all the selection criteria, including bootstrap photography of a leading landing-site candidate for Apollo 13 (Fra Mauro) and an opportunity to land next to a previously landed spacecraft (Surveyor III).

The Surveyor III site was located in a fairly distinct pattern of surface features which are necessary to the crew's ability to recognize and redesignate to the target. Figure 3-24 illustrates how effectively the goal of landing near the Surveyor was achieved.

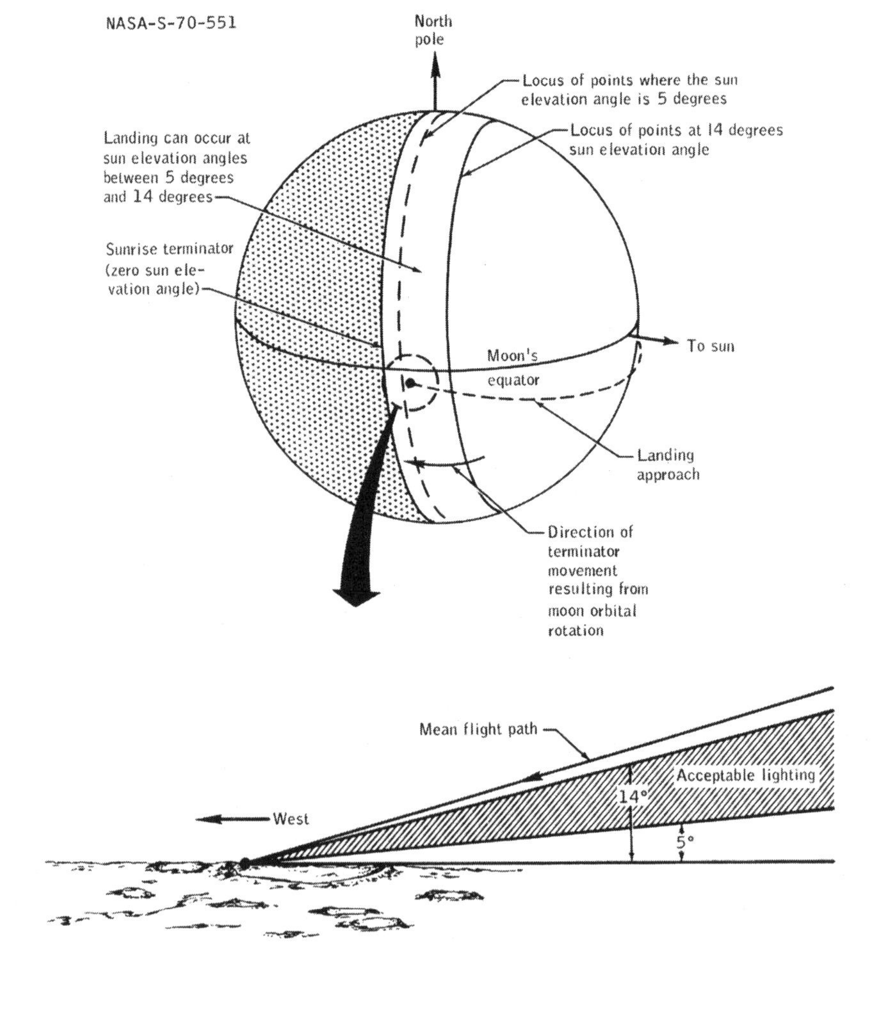


Figure 4-1 Sun elevation angle for lunar landing.

### 4.1.2 Launch Window Factors

There are a number of considerations which determine the unique time periods, called launch windows, from which a lunar landing mission can be flown. These considerations include illumination conditions at launch, launch azimuth, translunar injection geometry, sun elevation angle at the lunar landing site, illumination conditions at earth landing, and the number and location of lunar landing sites.

The time of lunar landing is essentially determined by the location of the lunar landing site and by the acceptable range of sun elevation angles (fig. 4-1). The range of acceptable sun elevation angles is from 5 to 14 degrees and in a direction from east to west. Under these conditions, visible shadows of craters aid the crew in recognizing topographical features. When the sun angle approaches the descent angle, the mean value of which is 16 degrees, visual resolution is degraded by a "washout" phenomenon where backward reflectance is high enough to eliminate contrast. Sun angles above the flight path are not as desirable because shadows are not readily visible unless the sun is significantly outside the descent plane. In addition, higher sun angles (greater than 18 degrees) can be eliminated from consideration by planning the landing one day earlier where the lighting is at least 5 degrees. Because lunar sunlight incidence changes about 1/2-degree per hour, the sun elevation angle restriction establishes a 16-hour period, which occurs approximately every 29.5 days, when landing at a given site can be attempted. The number of earth-launch opportunities for a given lunar month is of course equal to the number of candidate landing sites.

The time of launch is primarily determined by the allowable variation in launch azimuth and by the location of the moon at spacecraft arrival. The spacecraft must be launched into an orbital plane that contains the position of the moon and its antipode at spacecraft arrival. A 34-degree launch-azimuth variation affords a launch period of approximately 4 hours 30 minutes. This period is called the daily launch window and is the time that the direction of launch is within the required range to intercept the moon.

Two launch windows occur each day; one is available for a translunar injection out of earth orbit in the vicinity of the Pacific Ocean and the other in the vicinity of the Atlantic Ocean. The injection opportunity over the Pacific Ocean is normally preferred because it usually permits a daytime launch.

## 4.2 DESCENT GUIDANCE AND CONTROL

While the lunar landing procedures and profile were generally similar to those of Apollo 11, the landing was intended to be a precision operation and a number of changes were incorporated primarily to reduce landing point dispersions. To eliminate related orbit perturbations, a soft undocking was performed with the spacecraft oriented radially with respect to the lunar surface. Also, physical separation of the spacecraft was performed using the service module reaction control system, and the lunar module 360-degree yaw maneuver and active stationkeeping activities were deleted. Because the landing point designator was to be used during the final stages of descent to facilitate manual redesignation of the target, a calibration was performed by sighting on a star at the elevation angle for which the descent trajectory was designed. To minimize the effect of accelerometer bias errors, the residuals following descent orbit insertion were not trimmed but were reported to the ground to be accounted for in a subsequent state vector update. The pitch-attitude drift check, which was performed on Apollo 11 by having the computer automatically point the telescope at the sun, was not required for Apollo 12 because a more accurate drift check was made prior to undocking. The more westerly landing site for Apollo 12 provided additional time between acquisition of signal and powered descent initiation; therefore, a state vector update could be made based on the previous revolution tracking and the confirmed descent orbit insertion residuals. In addition to this data-link update, the capability for manually updating the landing-site coordinates was provided, based on a voice update from the ground after starting powered descent. Descent was initiated in a face-up attitude; therefore, a 180degree yaw maneuver was not required after ignition. Because of this face-up attitude, no landing point altitude check, downrange position check, or horizon attitude check were performed.

Flight plan changes from Apollo 11 after touchdown included two rendezvous-radar tracking passes of the command module: one immediately after touchdown and the other just prior to ascent. In addition, the primary and abort guidance systems were powered down on the surface to conserve power.

### 4.2.1 Preparation for Powered Descent

Table 4-I contains a sequence of events for the lunar landing phase. System power-up and primary and abort guidance system alignments and drift checks all proceeded according to plan. An accelerometer bias update was performed as scheduled. Undocking and separation were also nominal, and the post-separation optical alignment of the inertial measurement unit indicated drifts well within allowable limits. Descent orbit insertion was reported on time with the following velocity residuals:

| Axis | Descent orbit insertion velocity residuals, ft/sec | |
|---|---|---|
| | Primary guidance | Abort guidance |
| X | 0 | 0.3 |
| Y | 0.2 | 0.1 |
| Z | -0.6 | -0.6 |

The Doppler residuals measured on the ground at acquisition of signal following descent orbit insertion indicated a downrange error of 4400 feet, and the initial output of the Network powered flight processor indicated a downrange error of 4200 feet. Therefore, a downrange landing point correction of 4200 feet was transmitted to the crew and inserted into the guidance computer approximately 1.5 minutes after ignition for powered descent.

**TABLE 4-I.- POWERED DESCENT SEQUENCE OF EVENTS**

| Time, hr:min:sec | Event |
|---|---|
| 110:00:28 | Braking phase program (P63) entered |
| 110:02:25 | Braking phase program (P63) exited |
| 110:13:39 | Start abort guidance system initialization |
| 110:14:37 | Abort guidance system initialization completed |
| 110:14:41 | Request rendezvous parameter display (Verb 83) called |
| 110:15:23 | Request rendezvous parameter display (Verb 83) terminated |
| 110:16:29 | Coupling display unit zero started |
| 110:16:45 | Coupling display unit zero completed |
| 110:20:03 | Display keyboard assembly blank (time to ignition - 35) |
| 110:20:08 | Average-g on (time to ignition -29.9) |
| 110:20:31 | Ullage (time to ignition -7.5) |
| 110:20:33 | Enable engine (Verb 99 ) |
| 110:20:37 | Ignition permitted |
| 110:20:38 | Ignition |
| 110:21:05 | Throttle up |
| 110:22:03 | Landing site correction (Noun 69) initiated |
| 110:22:27 | Landing site correction (Noun 69) entered |
| 110:24:00 | Landing radar altitude lock |
| 110:24:04 | Landing radar velocity lock |
| 110:24:09 | Permit landing radar updates (Verb 57) entered |
| 110:24:25 | State-vector update allowed |
| 110:24:31 | Permit landing radar updates (Verb 57) exited |
| 110:26:08 | Abort guidance system altitude update |
| 110:26:24 | Velocity update initiate |
| 110:26:39 | X-axis override inhibited |
| 110:27:01 | Throttle recovery |
| 110:27:26 | Abort guidance system altitude update |
| 110:29:11 | Approach phase (P64) entered |
| 110:29:14 | Landing point designator enabled |
| 110:29:18 | Landing radar antenna position 2 |
| 110:29:20 | Abort guidance system altitude update |
| 110:29:44 | Redesignation right |
| 110:29:47 | Landing radar low scale |
| 110:30:02 | Redesignation long |
| 110:30:06 | Redesignation long |
| 110:30:12 | Redesignation right |
| 110:30:30 | Redesignation short (2) |
| 110:30:42 | Redesignation right |
| 110:30:46 | Attitude hold |

| | |
|---|---|
| 110:30:50 | Rate of descent landing phase (P66) entered |
| 110:31:18 | Landing radar data dropout |
| 110:31:24 | Landing radar data recovery |
| 110:31:27 | Landing radar data dropout |
| 110:31:37 | Landing radar data recovery |
| 110:32:00 | Landing radar data dropout |
| 110:32:04 | Landing radar data recovery |
| 110:32:35 | Engine off |
| 110:32:36 | Touchdown |

### 4.2.2 Powered Descent

The ignition sequence for powered descent was nominal and occurred on time. The desired landing site was approximately 5 miles south of the orbital plane; therefore, an initial roll angle of minus 4 degrees resulted as the spacecraft was steered to the left by descent guidance. Figure 4-2 (a) is an altitude-versus-altitude-rate profile for data from the primary and abort guidance systems and the tracking network, and figure 4-2 (b) is a plot of altitude and altitude rate-versus time for the primary guidance system. Figures 4-3 and 4-4 show similar comparisons of horizontal and lateral velocity. The data show close agreement between all sources and indicate excellent systems performance. Lateral velocity reached a maximum of 78 feet per second approximately 5 minutes after ignition. This large out-of-plane velocity resulted from the 5-mile crossrange steering required during descent. Figure 4-5 shows a comparison of the commanded thrust level versus horizontal velocity for the primary guidance system with that predicted by the preflight operational trajectory. The actual thrust command profile was below nominal because the 4200-foot update in landing position resulted in early throttle-down.

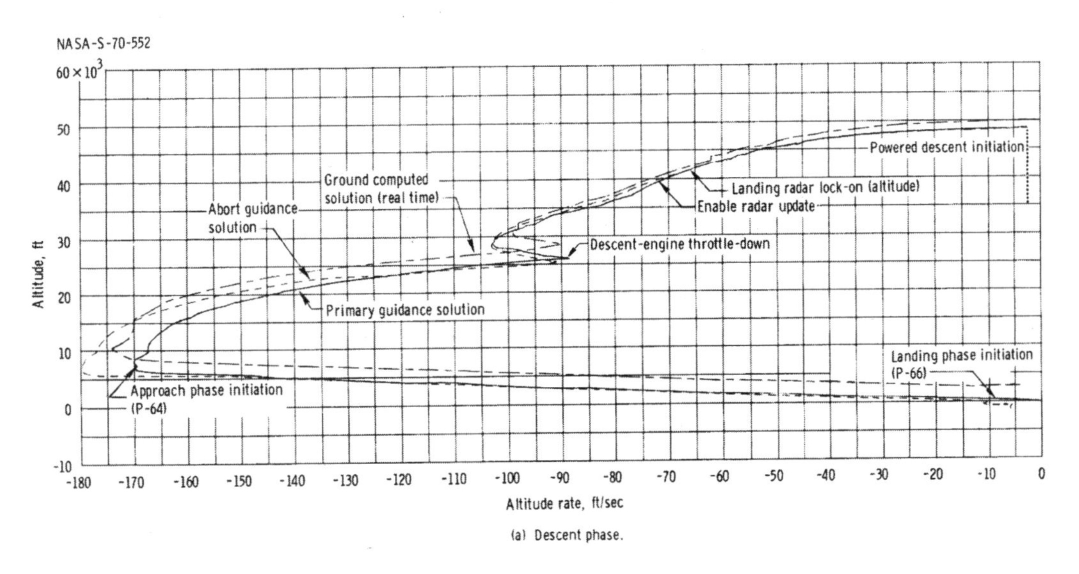


Figure 4-2 Comparison of altitude and altitude rate.

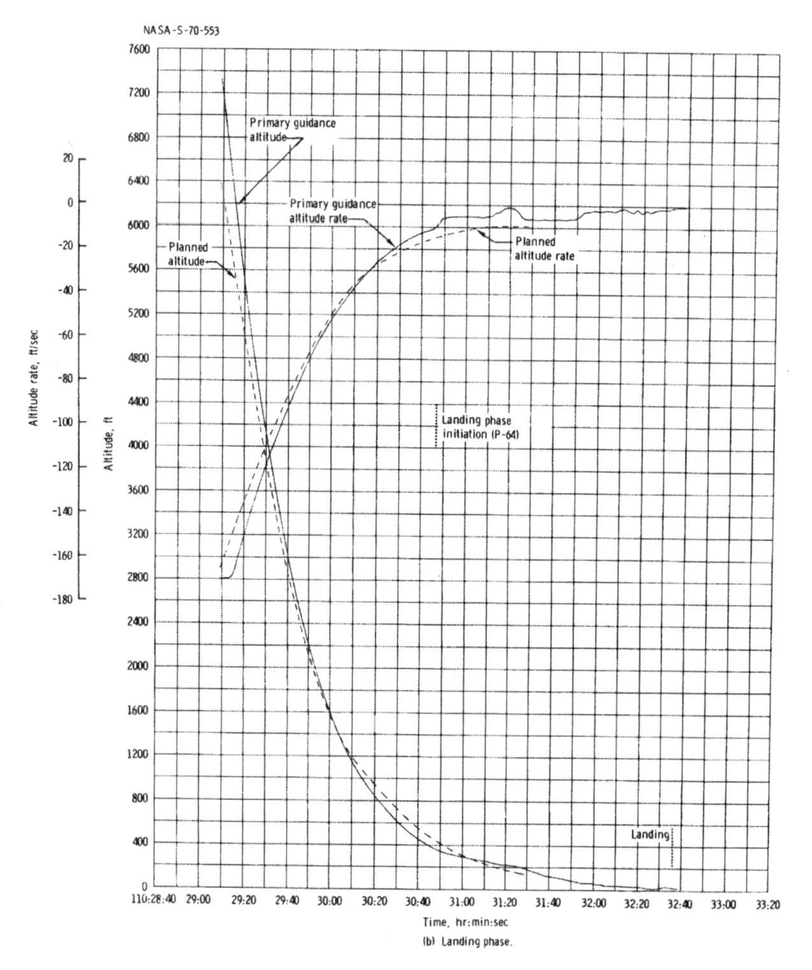


Figure 4-2 Concluded

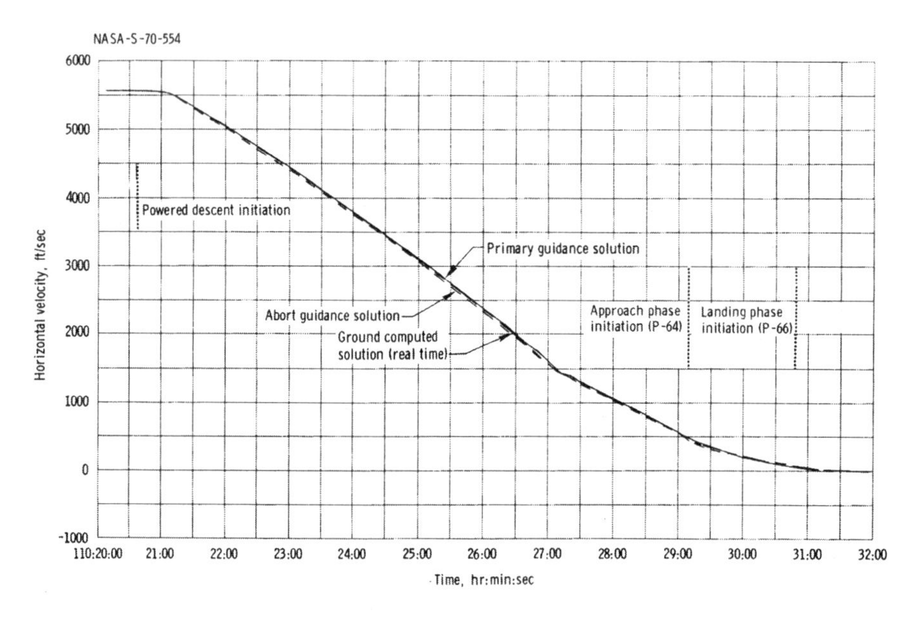


Figure 4-3 Horizontal velocity during descent

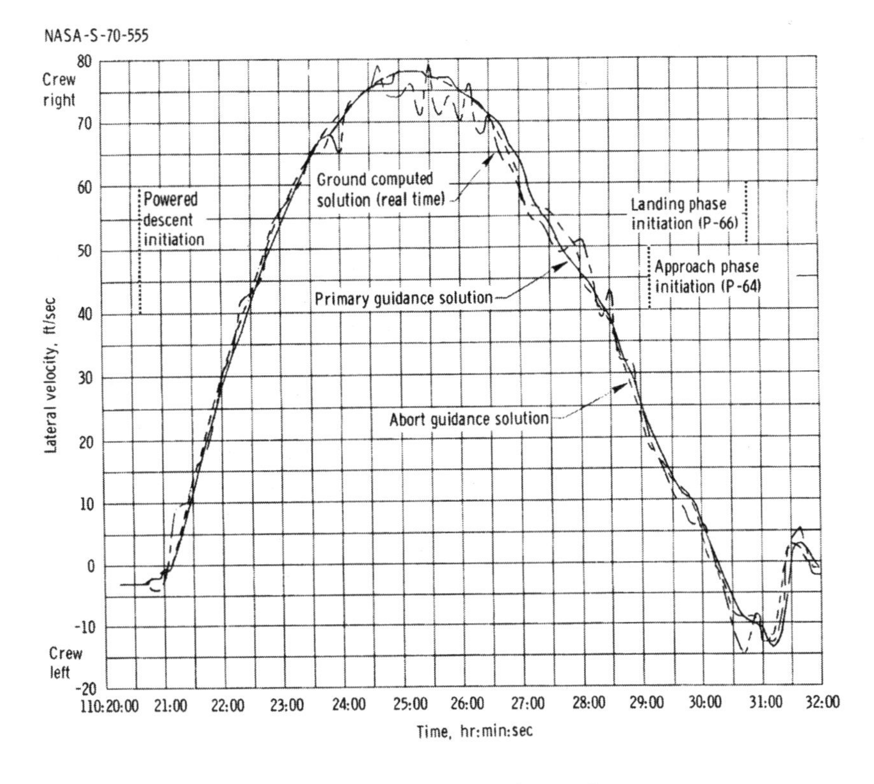


Figure 4-4 Lateral velocity during descent

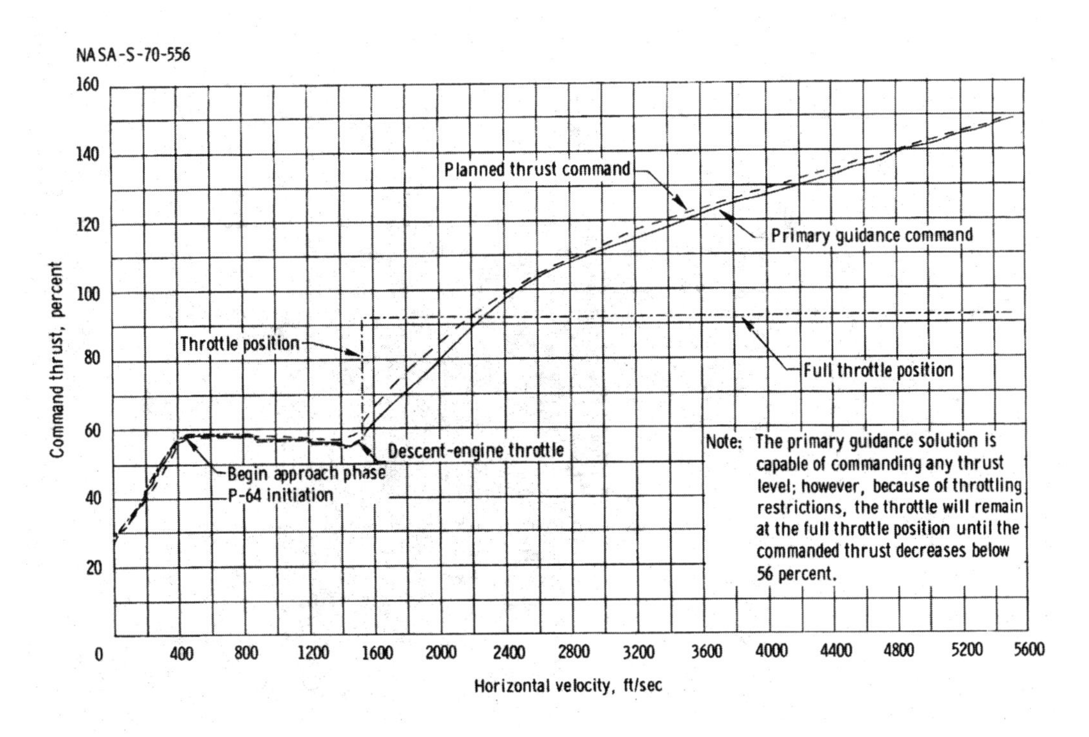


Figure 4-5 Comparison of percent commanded thrust and horizontal velocity.

Landing radar acquisition in altitude occurred at 41,438 feet and in velocity 4 seconds later at an altitude of 40,100 feet, which was well above that predicted before flight. Figure 4-6 contains the altitude difference time history between the altitude measured by the landing radar and that contained in the onboard guidance system. The initial difference of approximately 1700 feet converged to about 400 feet within 30 seconds after radar updates were enabled and to approximately 100 feet within 2 1/2 minutes. Radar data remained stable until at 80 seconds before touchdown the two rear velocity beams entered regions of zero Doppler. As expected, a limited degradation of altitude and velocity data existed from this point until touchdown.

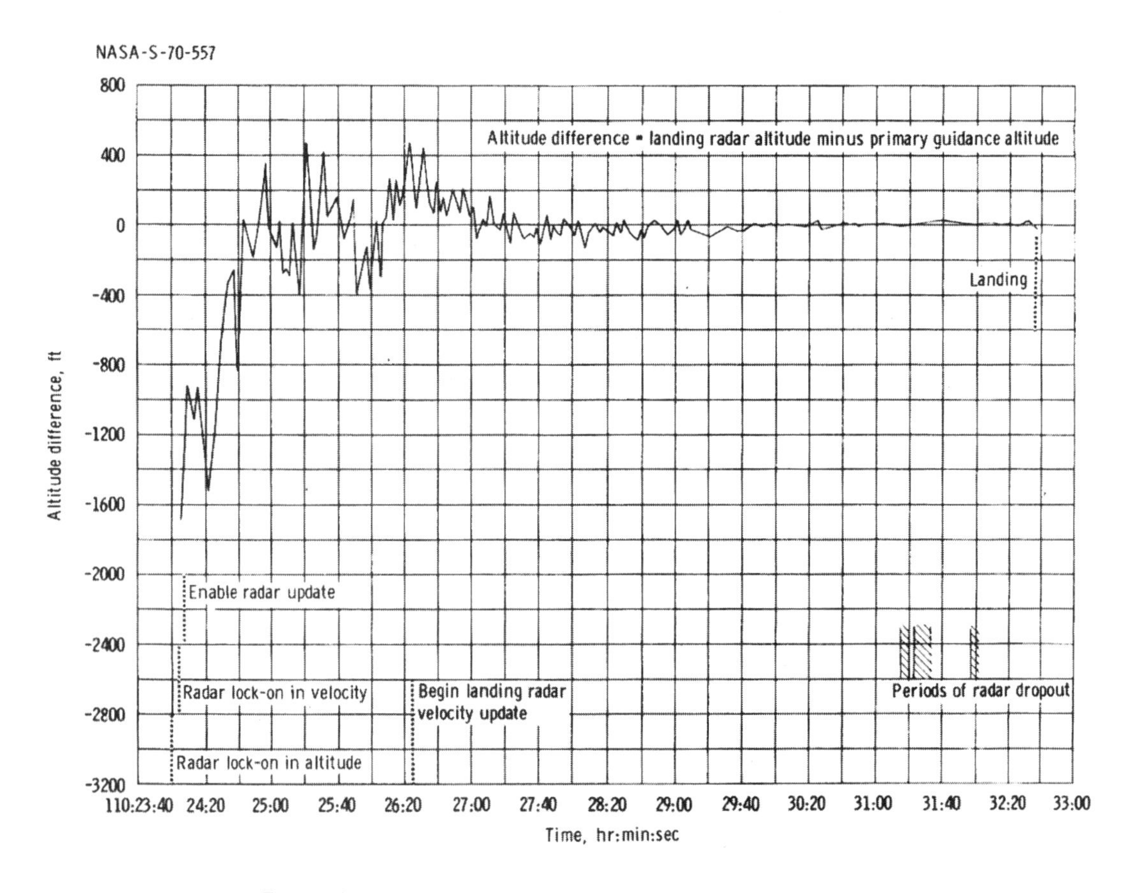


Figure 4-6 Altitude difference between radar and primary guidance

Figure 4-7 contains a time history of pertinent control system parameters during the powered descent phase. The dynamic response of the spacecraft was nominal throughout this phase, although the crew reported an unexpected amount of reaction control system activity. The following table indicates that reaction control propellant utilization was very close to that evident in preflight simulations of the automatic phases of descent.

Reaction control propellant used, lb

| Phase | Predicted | Actual |
|---|---|---|
| Braking | 15.2 | 15.7 |
| Approach | 16.9 | 16.3 |
| Landing | * | 60.3 |

*Nominal flight planning only accounts for automatic system usage.

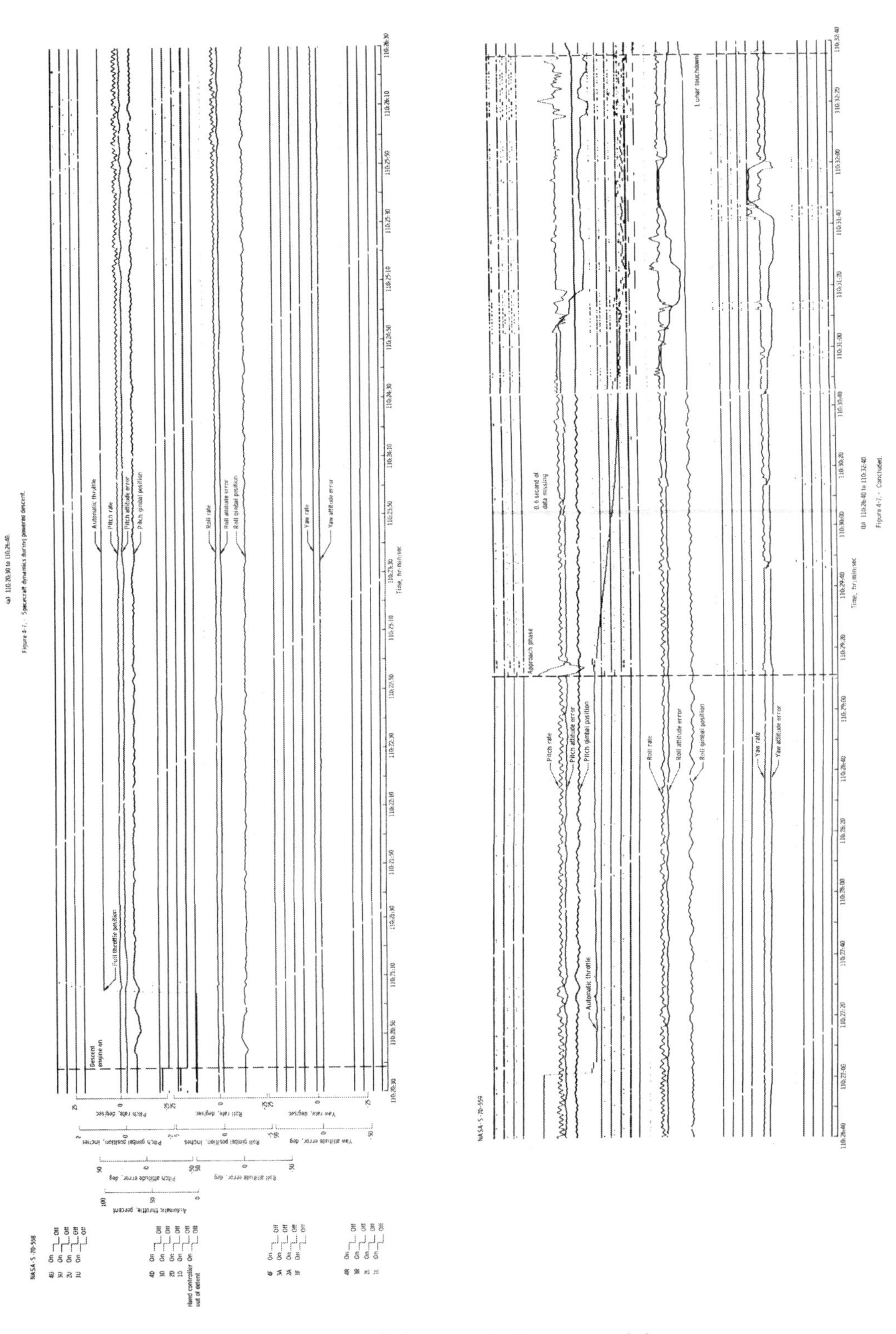

Figure 4-7 Spacecraft dynamics during powered descent.

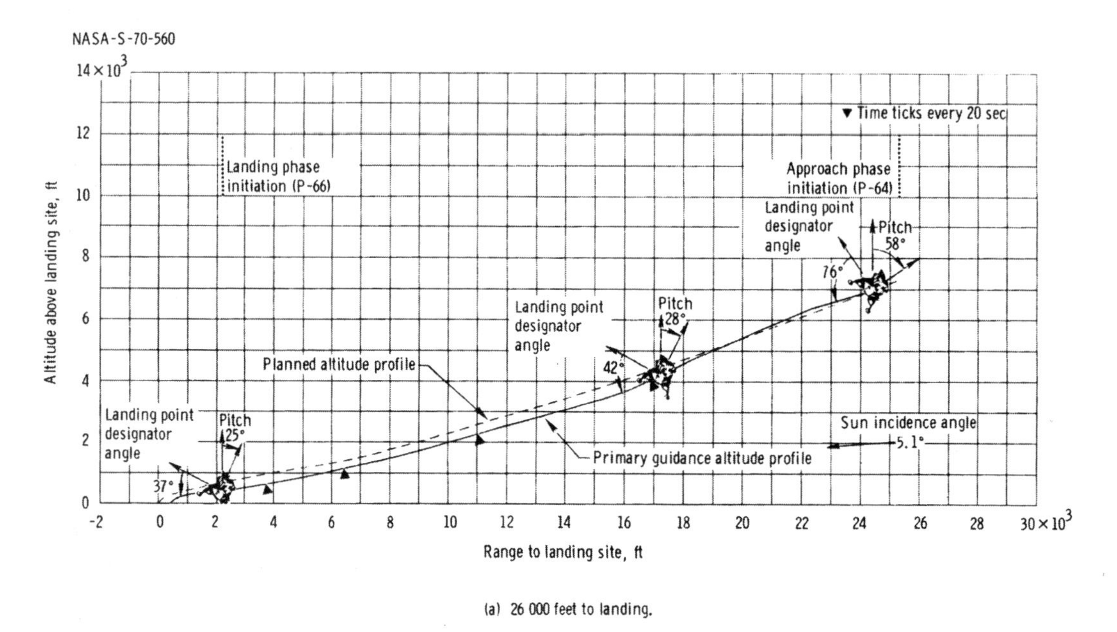


(a) 26 000 feet to landing.

Figure 4-8 Comparison of altitude and range from the landing site.

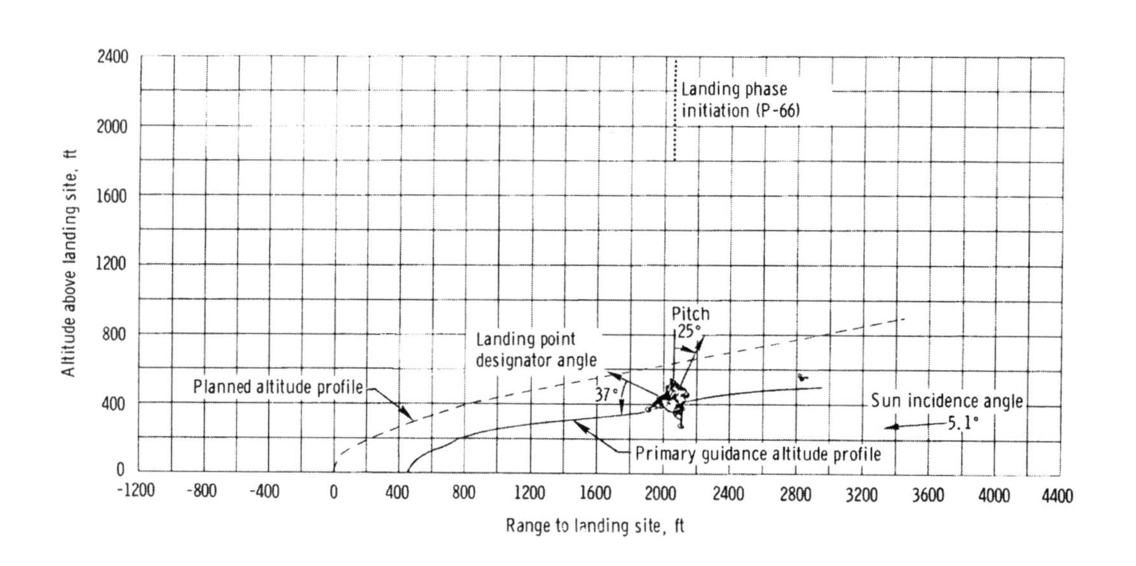


(b) 4200 feet to landing.

Figure 4-8 Concluded

(a) Training photograph.

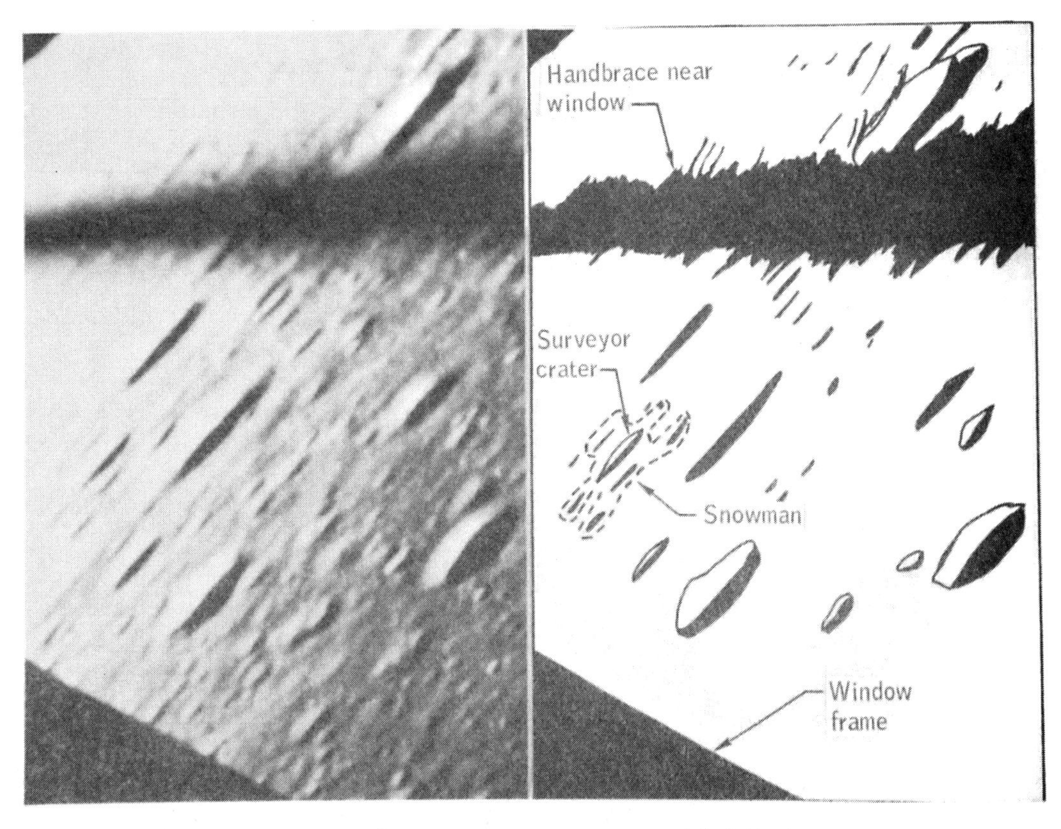


(b) Actual photograph.

(c) Artist's drawing.

Figure 4-9 Apollo 12 landing site

The automatic transition to the approach phase at high-gate (fig. 4-8) occurred at the near-nominal conditions of 6989 feet in altitude and 170 ft/sec in velocity. Following the pitchover maneuver, which was performed automatically to provide landing site terrain visibility, the computer began providing landing-point-designator elevation look angles. The crew reported that the displayed look angle was on target and that the series of craters in the configuration of a "snowman" was immediately visible (fig. 4-9). Figure 4-10 contains a time history of landing-point-designator look angles. Seven redesignations of the landing site were manually commanded by displacing the rotational hand controller out of detent in the desired direction. The effect of these control inputs on the landing point is indicated graphically and on the site map in figure 4-11. The total effect was to redefine the automatic target point 718 feet to the right and 361 feet downrange of the initial target. During final descent, the lunar module traveled approximately 1500 feet downrange, or about 400 feet less than the automatic target which existed after the seven manual redesignations.

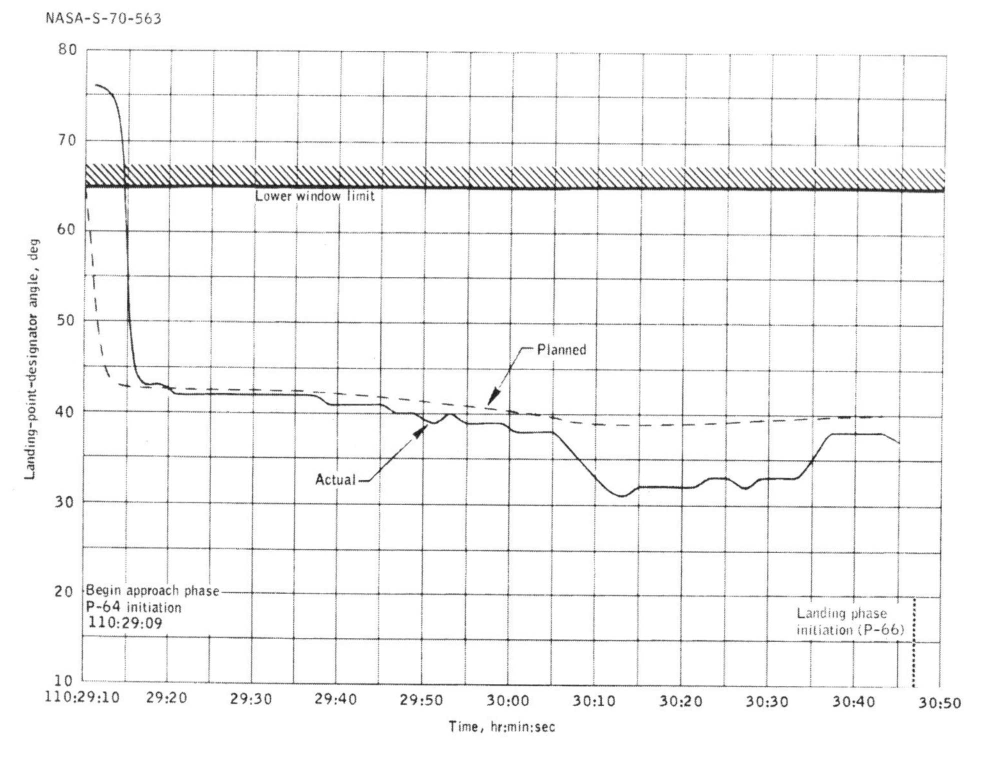


Figure 4-10 Comparison of landing point designator angle and time during approach phase.

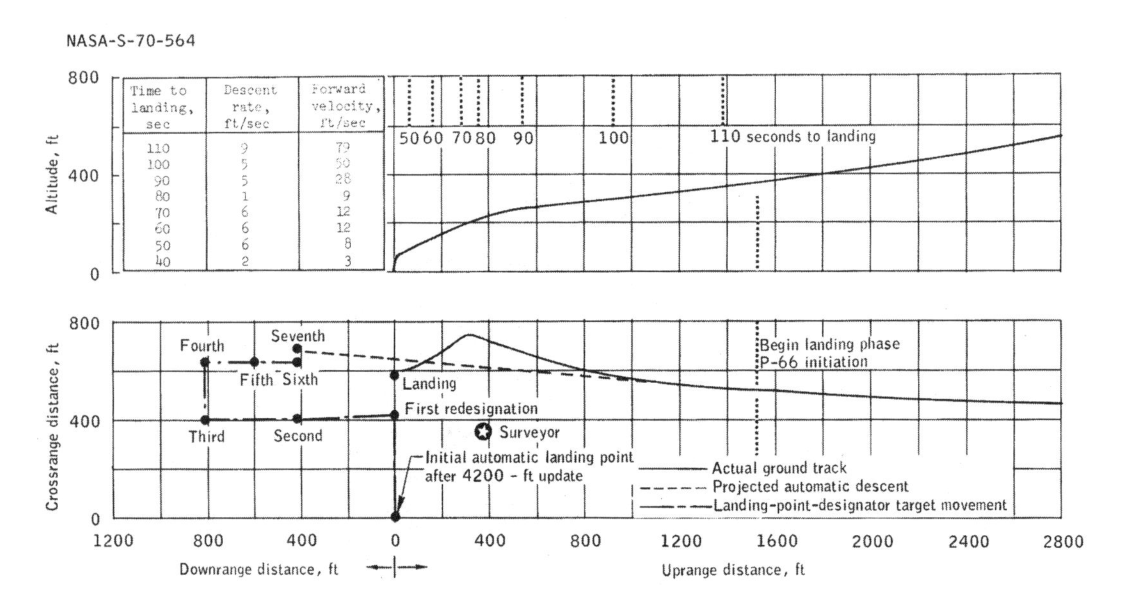


| Time to landing, sec | Descent rate, ft/sec | Forward velocity, ft/sec |
|---|---|---|
| 110 | 9 | 79 |
| 100 | 5 | 50 |
| 90 | 5 | 28 |
| 80 | 1 | 9 |
| 70 | 6 | 12 |
| 60 | 6 | 12 |
| 50 | 6 | 8 |
| 40 | 2 | 3 |

(a) Altitude and range from landing site.

Figure 4-11 Landing phase altitude and range histories

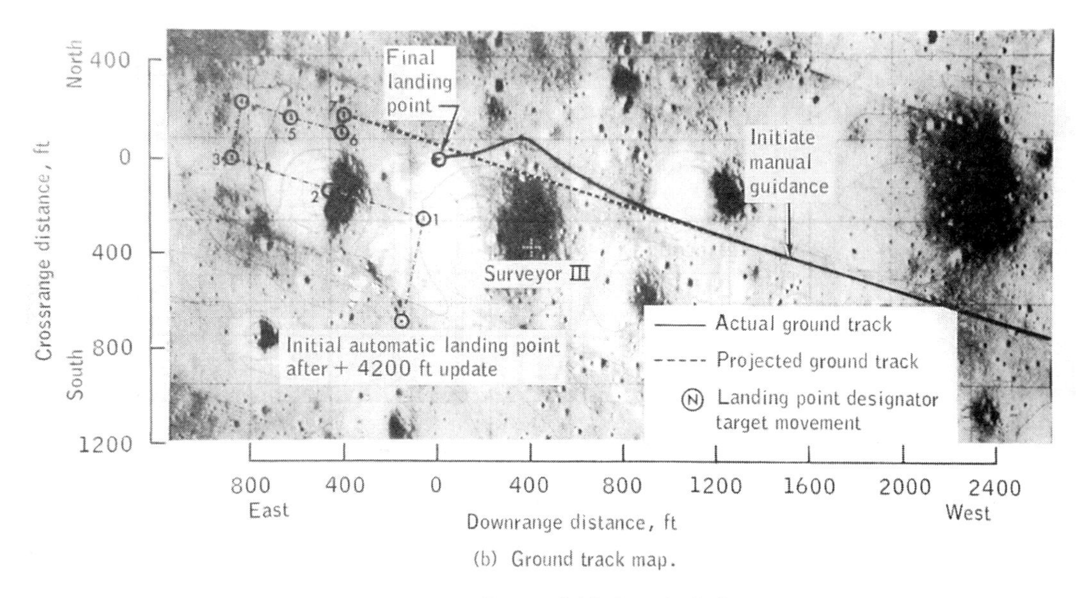


(b) Ground track map.

Figure 4-11 Concluded

The landing phase was performed manually, as expected, with an entry into the final-descent computer program (P66) at approximately 368 feet in altitude and at a descent rate of minus 8.8 ft/sec. The Commander reported that a check of the cross-pointers was made during this period and that zero velocity readings on the downrange and crossrange indicators was obtained on both the high- and low-sensitivity scales. The horizontal velocity measured by the primary guidance system is compared with altitude in figure 4-12, which indicates the descent was essentially vertical from the 50-foot altitude and that the horizontal velocity displayed was less than 1 ft/sec at times. The display is serviced by the computer every 0.25-second in 0.55-ft/sec steps. If the Commander's observation was made with an actual velocity of less than 1 ft/sec, it is possible that a near-zero reading could have existed. There are no data indications of abnormal hardware or software performance associated with the cross-pointers, and the pointers operated properly during ascent.

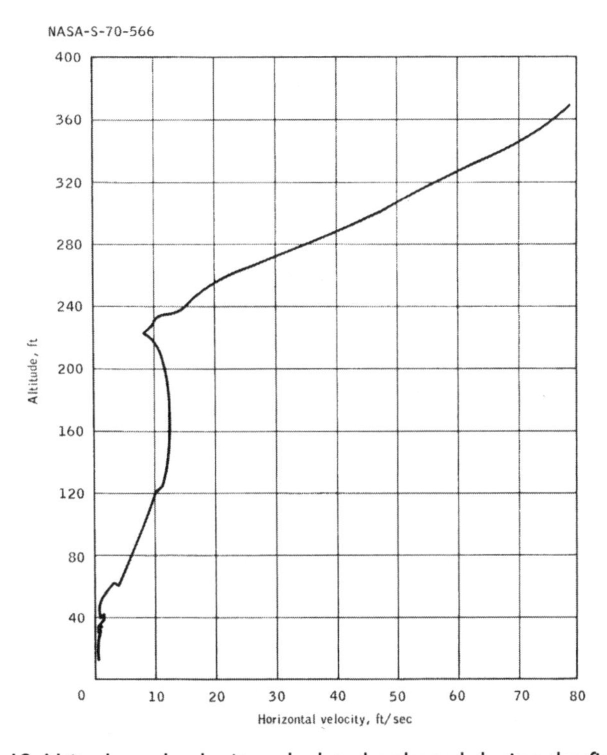


Figure 4-12 Altitude and velocity calculated onboard during the final descent phase.

Figure 6-1 contains a sequence of out-the-window photographs showing the effect of dust on visibility during the final phases. Section 4.3 contains a discussion and presentation of the actual landing site coordinates, and section 8.7 summarizes the descent propulsion system performance and operational margins.

### 4.2.3 Landing Dynamics

Figure 4-13 contains a time history of attitude rates near lunar touchdown, which occurred with first footpad contact at 110:32:36. The vehicle came to a stable rest within 1.5 seconds of this time. The descent engine stop button was activated approximately 1.3 seconds prior to first pad contact, and the engine thrust was consequently in a transient decay at the time surface contact occurred. The vertical velocity at the time the engine stop button was activated was approximately 0.4 ft/ sec downward and increased to about 3.2 to 3.5 ft/sec before first footpad contact. At the time of contact, the forward velocity was approximately 1.7 ft/sec, with a lateral velocity to the crew's left of about 0.4 ft/ sec. The final resting attitude, as viewed by the crew, was 3 degrees up in pitch and a 3.8-degree roll left, which indicates a surface slope of about 4 or 5 degrees downward to the left and rear of the crew. Pitch and roll attitudes at contact were approximately 3 degrees down and 1.4 degrees left, respectively. The primary spacecraft motion during landing was a pitching motion from the 3-degree pitch-down attitude to the final 3-degree pitch-up attitude, with a maximum pitch rate during this period of 19.5 deg/sec. This pitching motion was accompanied by a slight left roll and right yaw motion, with maximum rates on these axes of 7.8 and 4.2 degrees per second, respectively.

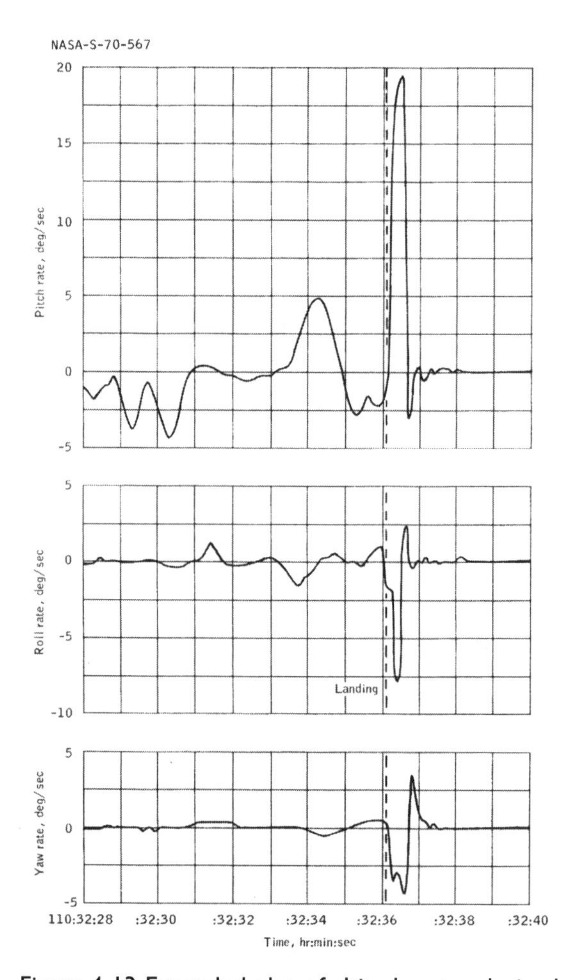


Figure 4-13 Expanded plot of altitude rates during landing

Digital computer simulations of the touchdown indicate that all primary strut strokes were less than 2.5 inches and secondary strut strokes were less than 4.5 inches. Maximum vertical and lateral accelerations during touchdown were less than 1 and 0.2 g, respectively. The coefficient of friction between the footpad and the lunar surface was approximately 0.4. The landing was very stable from a tipover standpoint, since the maximum angle between the spacecraft vertical axis and the local gravity vector did not exceed 4 degrees. The conclusions from the computer simulations of the landing dynamics are substantiated by crew comments and photographs of the landing gear and local surface.

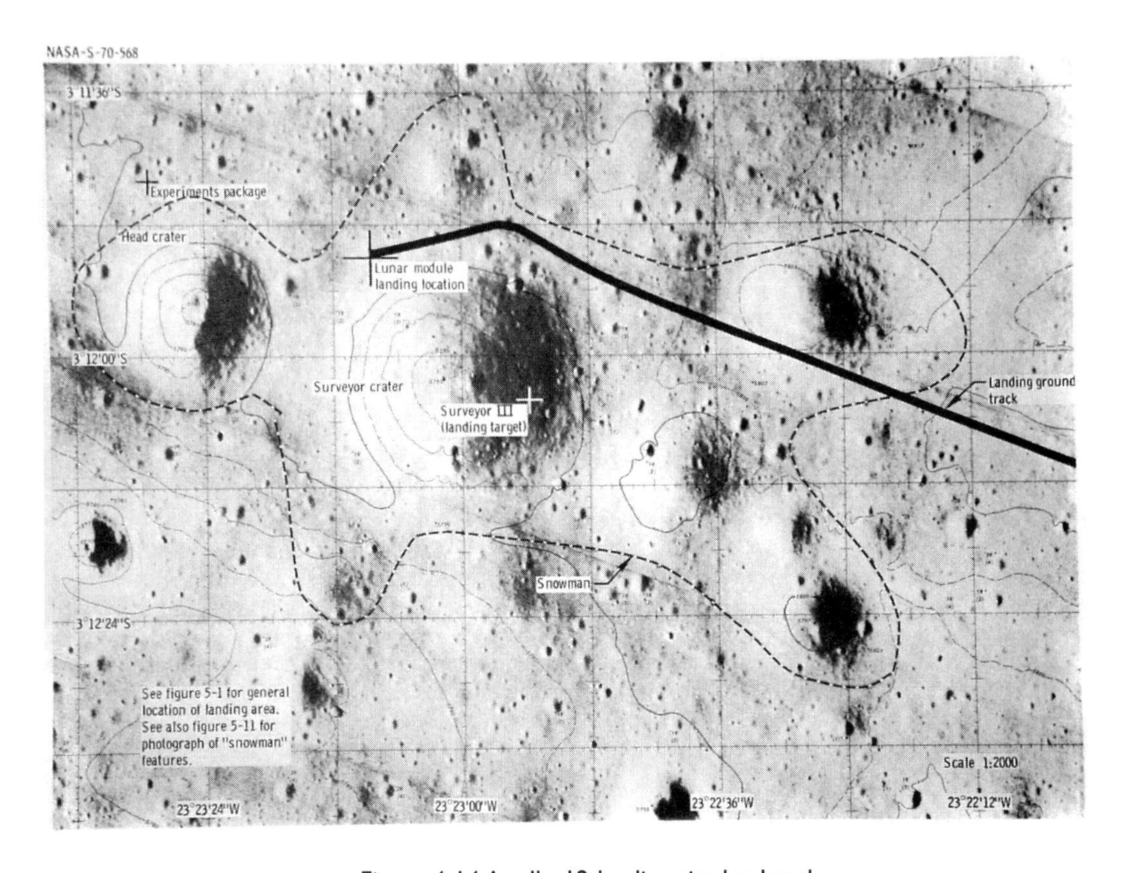


Figure 4-14 Apollo 12 landing site landmarks

### 4.3 LANDING SITE COORDINATES

Once the most valid reference map is chosen for a given landing site, the target coordinates and landing ellipse are given to trajectory analysts for preflight determination of spacecraft performance requirements and generation of reference trajectories. Prior to generation of the reference trajectories, the landing coordinates are converted into the inertial reference frame of the onboard guidance system through a reference-system transformation. The onboard targeting is therefore somewhat modified from the original coordinate reference to maintain consistency with onboard software. During the flight as tracking and navigation data become available, targeting coordinates may be further modified to account for known deficiencies in the lunar potential model and other constants. The location of the landing site relative to the lunar module, once it is separated from the command module, is computed in real time during lunar orbit, and the final targeting values are transmitted to the lunar module computer on the landing pass. The landing site position is biased from the preflight values to correct for errors in the location of both the landing site and the lunar module, based on lunar orbit navigation data. Therefore, it is not meaningful to compare stored landing coordinates with the actual site location because of the various transformations and targeting biases which have necessarily taken place. The entire real-time navigation and guidance operation, including ground-based computations and updates, proved the capability to perform a precision landing at a designated location.

Insofar as the landing site was concerned on Apollo 11, the only objective was to achieve a safe landing anywhere in the vicinity of the preselected landing area. For Apollo 12, however, considerable attention was devoted to achieving touchdown in close proximity to the targeted landing point. This preselected point was established coincident with the Surveyor III location, as shown in figure 4-14 and referenced to the Surveyor III Site Map (first edition, January 1968 ). Normal navigation uncertainties and guidance dispersions were expected to displace the actual automatic landing location sufficiently away from the Surveyor and the crater containing it that no landing hazard was presented the crew. In addition, if the descent path were exactly nominal, the crew could apply manual site redesignation in ample time to land outside the Surveyor crater. Actually, as discussed in the previous section, the unperturbed (automatic) descent trajectory was very close to nominal (170 feet south and 380 feet west of Surveyor), and the crew elected to over-fly the crater to the right side, eventually touching down very near its far rim. The final landing location, which was 535 feet from the Surveyor, was influenced by the preflight consideration that the landing occur outside a 500-foot radius of the target to minimize contamination of the Surveyor vehicle by descent engine exhaust and any attendant dust excitation.

The location of the actual touchdown point was first determined in real time from crew comments regarding surface features in the proximity of the vehicle. This determination was then confirmed from a variety of sources, including rendezvous radar data, ground tracking, onboard guidance parameters, and sextant sightings from lunar orbit. None of these sources, taken separately, are precise enough to establish within a few feet the location of the landing site with respect to known features.

The primary sources of information for locating the landing site during postflight analysis were the onboard sequence camera photographs (figs. 4-9 and 6-1) and triangulation from surface photography (for example, fig. 3-24 ). During preflight training, the crew used a series of craters, which approximated the shape of a "snowman" (fig. 4-9 ), to aid in their recognition of Surveyor crater during descent. The parts of this figure show first, the image used in preflight training exercises; second, the actual "snowman," as photographed during descent; and third, an artist's sketch to aid in locating the "snowman" from the actual photograph.

These information sources produced the actual landing site coordinates, as referenced to the Surveyor III Site Map (first edition, January 1968 ), of 3 degrees 11 minutes 51 seconds south latitude and 23 degrees 23 minutes 7.5 seconds west longitude. Other postflight data sources, including the best estimated trajectory and the reduced navigation data from the onboard guidance system, in general confirm this final landing location.

It should be noted that the stated coordinates are not valid for other reference maps because of variations in the grid coordinates from one map to another. That is, on larger scale maps in which the "snowman" and, in particular, Surveyor crater are visible, use of the reported landing site coordinates will not place the touchdown location in the same position relative to landing site features.

## 5.0 TRAJECTORY

The trajectory profile for this mission was similar to that for Apollo 11 , except for the inclusion of a non-free-return translunar profile and the deorbiting of the ascent stage after rendezvous. In addition, Apollo 12 had as an objective the demonstration of techniques for a precision lunar landing.

The analysis of the trajectory from lift-off to spacecraft/S-IVB separation was based on launch vehicle onboard data, as reported in reference 5, and from Network tracking data. After separation, the actual trajectory information was determined from the best estimated trajectory generated from tracking and telemetry data.

The earth and moon models used for the trajectory analysis are geometrically described as follows: (1 ) the geodetic earth model is a Fischer ellipsoid and the earth potential model is a fourth-order expansion which expresses the oblateness and other effects; and (2) the lunar potential model, new for this mission, describes the non-spherical potential field of the moon. This model, termed L1, is essentially the R2 model used previously but with an extra term added to permit improved determination and prediction of latitude and orbital period. The new L1 potential function is defined in a published revision to reference 6. Table 5-I is a listing of major flight events, and table 5-II defines the trajectory and maneuver parameters.

**TABLE 5-I . - SEQUENCE OF EVENTS**

Range zero - 16:22:00 G.m.t., Nov. 14, 1969

|  |  |
|---|---|
| Lift-off | 00:00:00.7 |
| S-IC outboard engine cutoff | 00:02:41.7 |
| S-IC/S-II separation | 00:02:42.4 |
| S-II engine ignition (command) | 00:02:44.2 |
| Launch escape tower jettison | 00:03:21.6 |

| | |
|---|---|
| S-II engine cutoff | 00:09:12.4 |
| S-IVB engine ignition (command) | 00:09:15.6 |
| S-IVB engine cutoff | 00:11:33.9 |
| Translunar injection maneuver | 02:47:23 |
| S-IVB/command and service module separation | 03:18:05 |
| Translunar docking | 03:26:53 |
| Spacecraft ejection | 04:13:01 |
| S-IVB separation maneuver | 04:26:41 |
| First midcourse correction | 30:52:44 |
| Lunar orbit insertion | 83:25:23 |
| Lunar orbit circularization | 87:48:48 |
| Undocking | 107:54:02 |
| First separation maneuver | 108:24:37 |
| Descent orbit insertion | 109:23:40 |
| Powered descent initiation | 110:20:38 |
| Lunar landing | 110:32:36 |
| First extravehicular egress | 115:10:35 |
| First extravehicular ingress | 119:06:38 |
| First lunar orbit plane change | 119:47:13 |
| Second extravehicular egress | 131:32:45 |
| Second extravehicular ingress | 135:22:00 |
| Lunar lift-off | 142:03:48 |
| Coelliptic sequence initiation | 143:01:51 |
| Constant differential height maneuver | 144:00:03 |
| Terminal phase initiation | 144:36:26 |
| Lunar orbit docking | 145:36:20 |
| Ascent stage jettison | 147:59:32 |
| Second separation maneuver | 148:04:31 |
| Ascent stage deorbit maneuver | 149:28:15 |
| Ascent stage impact | 149:55:16 |
| Second lunar orbit plane change | 159:04:46 |
| Transearth injection maneuver | 172:27:17 |
| Second midcourse correction | 188:27:16 |
| Third midcourse correction | 241:22:00 |
| Command module/service module separation | 244:07:20 |
| Entry interface | 244:22:19 |
| Landing | 244:36:25 |

**TABLE 5-11.- DEFINITION OF TRAJECTORY AND ORBITAL PARAMETERS**

| Trajectory Parameters | Definition |
|---|---|
| Geodetic latitude | Spacecraft position measured north or south from the earth's equator to the local vertical vector, deg |
| Selenographic latitude | Spacecraft position measured north or south from the true lunar equatorial plane to the local vertical vector, deg |
| Longitude | Spacecraft position measured east or west from the body's prime meridian to the local vertical vector, deg |
| Altitude | Perpendicular distance from the reference body to the point of orbit intersect, ft or miles; altitude above the lunar surface is referenced to the altitude of the landing site with respect to mean lunar radius |
| Space-fixed velocity | Magnitude of the inertial velocity vector referenced to the body-centered, inertial reference coordinate system, ft/sec |
| Space-fixed flight-path angle | Flight-path angle measured positive upward from the body-centered, local horizontal plane to the inertial velocity vector, deg |
| Space-fixed heading angle | Angle of the projection of the inertial velocity vector onto the local body-centered, horizontal plane, measured positive eastward from north, deg |
| Apogee | Maximum altitude above the oblate earth model, miles |
| Perigee | Minimum altitude above the oblate earth model, miles |

| | |
|---|---|
| Apocynthion | Maximum altitude above the moon model, referenced to landing site altitude, miles |
| Pericynthion | Minimum altitude above the moon model, referenced to landing site altitude, miles |
| Period | Time required for spacecraft to complete 360 degrees of orbit rotation, min |
| Inclination | Acute angle formed at the intersection of the orbit plane and the reference body's equatorial plane, deg |
| Longitude of the ascending node | Longitude where the orbit plane crosses the reference body's equatorial plane from below, deg |

For the first time, the S-IVB was targeted for a high-pericynthion free-return translunar profile, with the first major spacecraft maneuver intended to lower the resulting pericynthion altitude to approximately 60 miles. Upon execution of this maneuver, described in figure 5-1, the spacecraft was then intentionally placed on a non-free-return trajectory.

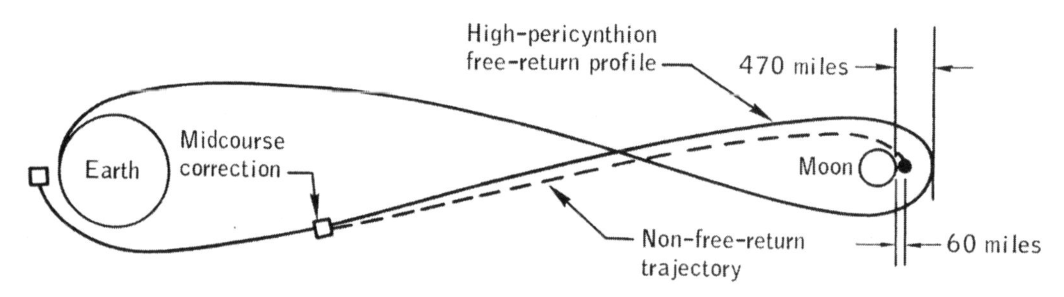


Figure 5-1 Hybrid non-free-return trajectory profile.

A free return profile, as used here, is a translunar trajectory that will achieve satisfactory earth entry within the reaction-control velocity correction capability. The major advantage of the new profile, termed a "hybrid" non-free-return trajectory, is the greater mission planning flexibility. This profile permitted a daylight launch to the planned landing site and a greater performance margin for the service propulsion system. Some of this margin was used to permit the two lunar orbit plane changes discussed later. The hybrid profile is constrained so that a safe return using the descent propulsion system can be made following a failure to enter lunar orbit. The trajectory parameters for the translunar injection and all spacecraft maneuvers are presented in table 5-III.

Following translunar injection, the pericynthion altitude of 470.7 miles was close to the real-time expected value. Because a state-vector error in the S-IVB guidance system was known to exist prior to translunar injection, the planned free-return conditions could not be achieved without an update of the guidance system. However, instead of performing an update, the projected pericynthion altitude was determined in view of the known error. Then, a new velocity change requirement for the midcourse correction to enter the desired non-free-return profile was determined. The actual velocity change of 61.8 ft/sec (table 5-IV) was about 0.1 ft/sec less than the real-time planned value and was applied at the second option point. No further translunar midcourse corrections were required. The maneuver to provide initial separation between the spacecraft and the S-IVB was accomplished for the first time on a lunar flight using the auxiliary propulsion system of the S-IVB. However, the final separation maneuver, performed as on previous lunar flights through S-IVB propulsive venting, did not place the S-IVB in a solar orbit, as planned, and the resulting orbit was a high-apogee ellipse (see section 13).

TABLE 5-III.- TRAJECTORY PARAMETERS

| Event | Ref. body | Time, hr:min:sec | Latitude, deg | Longitude, deg | Altitude, mile | Space-fixed velocity, ft/sec | Space-fixed flight-path angle, deg | Space-fixed heading angle, deg E of N |
|---|---|---|---|---|---|---|---|---|
| **Translunar Phase** | | | | | | | | |
| S-IVB second ignition | Earth | 2:47:22.7 | | | | | | |
| S-IVB second cutoff | Earth | 2:53:03.9 | 15.83N | 154.98W | 192.1 | 35 427 | 8.21 | 63.69 |
| Translunar injection | Earth | 2:53:14 | 15.83N | 154.98W | 192.1 | 35 427 | 8.21 | 63.69 |
| Command and service module/S-IVB separation | Earth | 3:18:04.9 | 28.82N | 79.57W | 3820.0 | 24 861 | 45.09 | 100.18 |
| Docking | Earth | 3:26:53.3 | 26.60N | 70.82W | 5337.4 | 22 534 | 49.89 | 105.29 |
| Spacecraft/S-IVB separation (ejection) | Earth | 4:13:00.9 | 18.50N | 58.63W | 12 504.5 | 16 447 | 60.93 | 114.52 |
| First midcourse correction | | | | | | | | |
| Ignition | Earth | 30:52:44.4 | 1.10S | 63.02W | 116 929.1 | 4 317 | 75.83 | 120.80 |
| Cutoff | Earth | 30:52:53.6 | 1.10S | 63.06W | 116 935.2 | 4 298 | 76.60 | 120.05 |
| **Lunar Orbit Phase** | | | | | | | | |
| Lunar orbit insertion | | | | | | | | |
| Ignition | Moon | 83:25:23.4 | 5.74N | 175.61E | 82.5 | 8 175 | -8.44 | 229.35 |
| Cutoff | Moon | 83:31:15.7 | 1.63S | 154.04E | 61.7 | 5 470 | -0.63 | 239.30 |
| Lunar orbit circularization | | | | | | | | |
| Ignition | Moon | 87:48:48.1 | 1.67S | 151.67E | 61.6 | 5 471 | -0.66 | 239.28 |
| Cutoff | Moon | 87:49:05 | 1.89S | 150.85E | 61.7 | 5 331 | .30 | 239.51 |
| Undocking | Moon | 107:54:02.3 | 13.52S | 86.96E | 63.0 | 5 329 | -0.03 | 267.25 |
| Separation | | | | | | | | |
| Ignition | Moon | 108:24:36.8 | 6.61S | 7.44W | 59.2 | 5 350 | -0.18 | 305.17 |
| Cutoff | Moon | 108:24:51.2 | 6.45S | 8.14W | 59.2 | 5 350 | -0.20 | 305.15 |
| Descent orbit insertion | | | | | | | | |
| Ignition | Moon | 109:23:39.9 | 6.64N | 172.21E | 60.5 | 5 343 | 0.17 | 234.81 |
| Cutoff | Moon | 109:24:08.9 | 6.29N | 170.76E | 61.5 | 5 268 | -0.02 | 234.85 |
| Powered descent initiation | Moon | 110:20:38.1 | 6.76S | 7.82W | 8.0 | 5 566 | -0.02 | 305.14 |
| Landing | Moon | 110:32:36.2 | 3.04S | 23.42W | -- | -- | -- | -- |
| Command and service module plane change | Moon | 119:47:13.2 | 14.01S | 77.68E | 62.2 | 5 334 | -0.07 | 269.27 |
| Coelliptic sequence initiation | | | | | | | | |
| Ignition | Moon | 143:01:51 | 5.16N | 164.68E | 51.5 | 5 310 | 0.06 | 234.43 |
| Cutoff | Moon | 143:02:32.1 | 4.65N | 162.64E | 51.5 | 5 355 | 0.02 | 234.29 |
| Terminal phase initiation | Moon | 144:36:26 | 14.57N | 128.99W | 44.5 | 5 382 | 0.05 | 257.93 |
| Docking | Moon | 145:36:20.2 | 14.53S | 46.98E | 58.1 | 5 357 | -0.04 | 284.29 |
| Command and service module/ascent stage separation | Moon | 148:04:30.9 | 1.40N | 43.34W | 59.9 | 5 347 | 0.15 | 304.19 |
| Ascent stage deorbit | | | | | | | | |
| Ignition | Moon | 149:28:14.8 | 14.32S | 62.86E | 57.6 | 5 362 | -0.12 | 272.27 |
| Cutoff | Moon | 149:29:36.9 | 14.47S | 58.62E | 57.4 | 5 177 | -0.27 | 275.90 |
| Ascent stage impact | Moon | 149:55:16.4 | 3.94S | 21.20W | -- | -- | -- | -- |
| Plane change | | | | | | | | |
| Ignition | Moon | 159:04:45.5 | 6.65S | 110.34E | 58.7 | 5 353 | -0.20 | 241.32 |
| Cutoff | Moon | 159:05:04.8 | 6.82S | 109.40E | 58.9 | 5 353 | -0.20 | 245.82 |
| Transearth injection | | | | | | | | |
| Ignition | Moon | 172:27:16.8 | 8.74N | 170.25W | 63.3 | 5 323 | -0.21 | 244.28 |
| Cutoff | Moon | 172:29:27.1 | 7.77N | 178.56W | 64.6 | 8 351 | 2.69 | 243.56 |
| **Transearth Coast Phase** | | | | | | | | |
| Second midcourse correction | | | | | | | | |
| Ignition | Earth | 188:27:15.8 | 15.88N | 137.80E | 180 031.1 | 3 036 | -78.44 | 91.35 |
| Cutoff | Earth | 188:27:20.2 | 15.88N | 137.78E | 180 028.9 | 3 036 | -78.40 | 91.36 |
| Third midcourse correction | | | | | | | | |
| Ignition | Earth | 241:21:59.7 | 14.78N | 92.40E | 25 059.0 | 12 083 | -68.54 | 96.00 |
| Cutoff | Earth | 241:22:05.4 | 14.78N | 92.38E | 25 048.3 | 12 085 | -68.55 | 96.01 |
| Command module/service module separation | Earth | 244:07:20.1 | 0.32N | 117.25E | 1 949.5 | 29 029 | -36.45 | 105.92 |

TABLE 5-IV.- TRANSLUNAR MANEUVER SUMMARY

| Maneuver | System | Ignition time, hr:min:sec | Firing time, sec | Velocity change, ft/sec | Resultant pericynthion conditions | | | | |
|---|---|---|---|---|---|---|---|---|---|
| | | | | | Altitude, miles | Velocity, ft/sec | Latitude, deg | Longitude, deg | Arrival time, hr:min:sec |
| Translunar injection | S-IVB | 2:47:22.7 | 341.3 | 10 515.0 | 280.2 | 7595 | 29.732S | 169.111E | 83:44:04.4 |
| Command and service module/S-IVB separation | Reaction control | 3:18:04.9 | | | | | | | |
| Spacecraft/S-IVB separation | S-IVB auxiliary propulsion system | 4:26:41.1 | 80.0 | | | | | | |
| First midcourse correction | Service propulsion | 30:52:44.4 | 9.2 | 61.8 | 65.1 | 8234 | 0.7N | 161.968E | 83:28:38.8 |

**TABLE 5-V.- LATITUDE TARGETING SUMMARY**

Landing site latitude on the landing revolution, deg

| | Apollo 10 | Apollo 11 | Apollo 12 |
|---|---|---|---|
| Desired | 0.691 north | 0.691 north | 3.037 south |
| Actual | 0.354 north | 0.769 north | 2.751 south |
| Error | 0.337 south | 0.078 north | 0.286 north |

The navigation data obtained during lunar orbit in preparation for descent was consistent with that of Apollo 10 and 11, but the projected landing-site latitude targeting was in greater error than that used for Apollo 11. Table 5-V shows that this error was of the same order as that experienced in Apollo 10 (0.286 versus 0.337 degree). Although not large, this error was compensated for in the final powered descent targeting. The 0.286 degree latitude error resulted from three primary sources. The first was the translunar navigation and lunar orbit insertion maneuver execution errors which contributed 0.039 degree. The second was due to an error in the landing site location which was discovered through command module optical tracking. The landing site was found to be 0.047 degree south of the prelaunch estimate. The third and largest was due to an error in the lunar potential model which failed to account properly for the lunar orbit motion. This source contributed 0.20 degree. A revised landing site location was also transmitted to the lunar module guidance computer soon after powered descent initiation (section 4.2.2) to correct for a 4200-foot downrange error which had been observed from ground tracking data. The more westerly landing site, as compared to Apollo 11, permitted sufficient time for acquisition and processing of later trajectory information just before descent so that these last-minute updates in the state vector and landing site location could be made, a procedure which is largely responsible for the precision with which the landing was performed. As in Apollo 10 and 11, the deficiencies in orbit prediction which are inherent in both the R2 and the new L1 potential models were accounted for through biasing of the targeting for lunar orbit insertion and circularization. The additional term which differentiates the L1 from the R2 potential function greatly improves the prediction accuracy of orbital period, a capability which permits return to a one-pass fit technique, as used in Apollo 8 and 10 (ref. 7 and 8). This change provides greater operational flexibility in ground tracking during lunar orbit coast and in the target updates prior to landing. Also, as in Apollo 11, the orbit was deliberately made noncircular to account for expected perturbations in lunar gravity such that the orbit would be more nearly circular during the rendezvous.

The descent, ascent, and rendezvous profiles were similar to those for Apollo 11, except that the landing point was changed. The descent operation is described in detail in section 4.2. Tracking data prior to undocking showed the ground track to be about 5 miles north of the intended landing site as a result of orbit-plane prediction uncertainties. A correction was combined with the powered descent maneuver to remove this discrepancy. The landing, as shown in figure 4-11, occurred within 535 feet of the Surveyor, at 3 degrees 11 minutes 51 seconds south latitude and 23 degrees 23 minutes 7.5 seconds west longitude (section 4.3), as referenced to the Surveyor III Site Map (lst ed., Jan. 1968).

Two plane changes were performed by the command and service module. The first was accomplished prior to lunar module ascent to accomodate normal movement of the lunar module out of the initial lunar-orbit plane resulting from the moon's rotation during the extended lunar stay. In the thirty-sixth lunar orbit revolution, the second plane change maneuver was conducted to permit photography of the landing areas and approach paths for future candidate landing sites. Both service propulsion maneuvers were nominal, with resultant errors less than 1 ft/sec. A summary of the lunar orbit maneuvers is shown in table 5-VI.

Lunar module ascent was nominal, except for a 1.2-second overburn caused by a late positioning of the engine-arm switch which inhibited the automatic cutoff signal. The relatively large residuals were subsequently nulled by the crew, and the rendezvous sequence which followed was nearly nominal (table 5-VII). Onboard solutions agreed closely with those computed in the command module and by the ground (table 5-VII).

The ascent stage was deorbited after jettison for a planned lunarsurface impact. A planned 200-ft/sec velocity change was provided by burning the remaining propellants through the reaction control system. The spacecraft impacted approximately 40 miles east-southeast of the Apollo landing site (fig. 5-2), as compared with an intended distance of 5 miles, primarily because of a 2-second overburn (5 ft/sec).

## TABLE 5-VI.- LUNAR ORBIT MANEUVER SUMMARY

| Maneuver | System | Ignition time, | firing time, | Velocity change, | Resultant orbit Apocynthion, | Pericynthion, |
|---|---|---|---|---|---|---|
| | | hr:min:sec | sec | ft/sec | miles | miles |
| Lunar orbit insertion | Service propulsion | 83:25:23.4 | 352.3 | 2889.5 | 168.8 | 62.6 |
| Lunar orbit circularization | Service propulsion | 87:48:48.1 | 16.9 | 165.2 | 66.1 | 54.3 |
| Command module/lunar module separation | Command module reaction control | 108:24:36.8 | 14.4 | 2.4 | 63.5 | 56.3 |
| Descent orbit insertion | Descent propulsion | 109:23:39.9 | 29.0 | 72.4 | 60.6 | 8.1 |
| Powered descent initiation | Descent propulsion | 110:20:38.1 | 717.0 | — | — | — |
| First lunar orbit plane change | Service propulsion | 119:47:13.2 | 18.2 | 349.9 | 62.5 | 57.6 |
| Lunar orbit insertion | Ascent propulsion | 142:10:59.9 | 423.2 | 6057.0 | 46.3 | 8.8 |
| Coelliptic sequence initiation | Lunar module reaction control | 143:01:51 | 41.1 | 45.0 | 51.0 | 41.5 |
| Constant differential height | Lunar module reaction control | 144:00:02.6 | 13.0 | 13.8 | 44.4 | 40.4 |
| Terminal phase initiation control | Lunar module reaction | 144:36:26 | 26.0 | 29.0 | 60.2 | 43.8 |
| Terminal phase finalization | Lunar module reaction control | 145:19:29.3 | 38.0 | 40.0 | 62.3 | 58.3 |
| Final separation control | Service module reaction | 148:04:30.9 | 5.4 | 1.0 | 62.0 | 57.5 |
| Lunar module deorbit control | Lunar module reaction | 149:55:16.4 | 82.1 | 196.3 | — | — |
| Second lunar orbit plane change | Service propulsion | 159:04:45.5 | 19.2 | 381.8 | 64.7 | 56.8 |

TABLE 5-VII.- RENDEZVOUS MANEUVER SOLUTIONS

| Maneuver | Lunar module | | | | Real-time nominal | | Command module guidance[a] | | Actual | |
|---|---|---|---|---|---|---|---|---|---|---|
| | Primary guidance | | Abort guidance | | | | | | | |
| | Time, hr:min:sec | Velocity, ft/sec | Time, hr:min:sec | Velocity, ft/sec | Time, hr:min:sec | Velocity, ft/sec | Time, hr:min:sec | Velocity, ft/sec | Time, hr:min:sec | Velocity, ft/sec |
| Coelliptic sequence initiation | 143:01:51 | 45.3 posigrade | 143:01:51 | 46.1 posigrade | 143:01:51 | 49.0 posigrade | 143:01:51 | 44.9 posigrade | 143:01:51 | 51.6 posigrade 0.1 south 0.3 down |
| Constant differential height | 144:00:02 | 10.2 retrograde 9.3 down | 144:00:02 | 9.4 retrograde 13.5 down | 143:59:53 | 2.3 down | 144:00:02 | 10.3 retrograde 0.4 south 7.8 down | 144:00:02 | 10.1 retrograde 9.1 down |
| Terminal phase initiation | 144:36:29 | 25.9 posigrade 1.5 south 11.9 down | 144:35:33 | 28.2 posigrade 1.7 south 10.9 down | 144:38:00 | 22.2 posigrade 0.1 south 10.9 down | 144:36:57 | 25.5 posigrade 1.7 south 10.9 down | 144:36:39 | 25.8 posigrade 1.4 south 11.1 down |
| First midcourse correction | 144:51:29 | 0.5 retrograde 2.0 up | 144:51:29 | 3.8 retrograde 0.3 north 4.6 down | | 0.0 | 144:51:29 | 1.6 retrograde 0.1 north 5.3 down | 144:51:29 | (b) |
| Second midcourse correction | 145:06:29 | 0.9 retrograde 0.3 south 0.7 down | (c) | (c) | | 0.0 | 145:06:29 | 0.1 retrograde 0.3 north 1.6 up | 145:06:29 | (b) |

[a] For lunar module execution; midcourse solutions obtained from VHF ranging data only (tracking light failed).
[b] Data not available because of moon occultation.
[c] Solution not obtained.

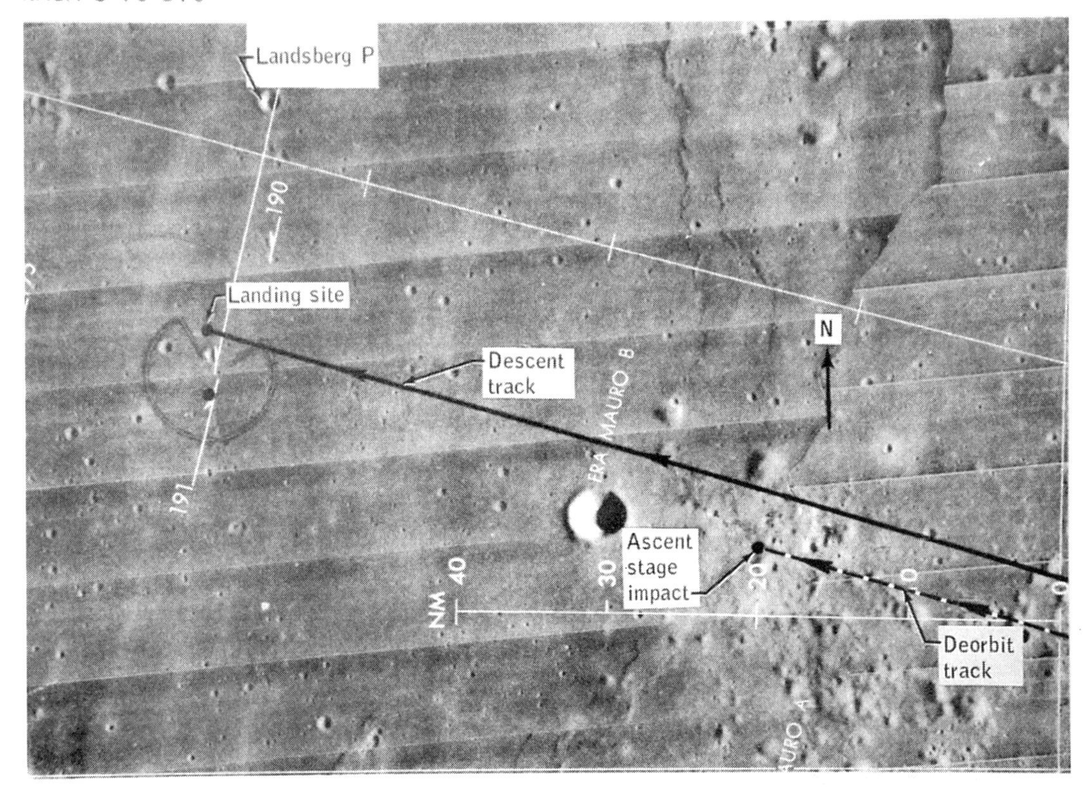


Figure 5-2 Preliminary landing and impact locations.

After transearth injection (table 5-VIII) and two subsequent midcourse corrections, the second at 3 hours prior to entry, entry was performed as planned. Entry parameters are listed in table 5-IX. The landing was within 2 miles of the intended location and occurred at 15 degrees 46.6 minutes south latitude and 165 degrees 9 minutes west longitude, as determined from the recovery ship.

Following separation from the command module, the service module reaction control system was fired to depletion. Based on stable service module attitudes during this firing, sufficient velocity change capability existed in the reaction-control-system to cause the service module to skip out into a high-apogee orbit. There was no radar or aircraft coverage planned for the service-module jettison and separation sequences. However, if the service module had skipped out as expected, it would probably have been visible to tracking stations which were alerted as to its expected position. No radar acquisition was made and no visual sightings by the crew or recovery personnel were reported. Therefore, as in previous missions, it is believed that the service module became unstable during the depletion firing and did not execute the velocity change required to skip out. Instead, the service module probably entered the atmosphere and impacted before detection.

## TABLE 5-VIII.- TRANSEARTH MANEUVER SUMMARY

| Event | System | Ignition time, hr:min:sec | Firing time, sec | Velocity change, ft/sec | Flight-path angle, deg | Resultant entry interface condition | | | Arrival time, hr:min:sec |
|---|---|---|---|---|---|---|---|---|---|
| | | | | | | Velocity, ft/sec | Latitude, deg | Longitude, deg | |
| Transearth injection | Service propulsion | 172:27:16.8 | 130.3 | 3042.0 | -7.24 | 36 116 | 13.55S | 172.11E | 244:21:49.3 |
| Second midcourse correction | Service module reaction control | 188:27:15.8 | 4.4 | 2.0 | -6.42 | 36 116 | 13.815 | 173.68E | 244:22:10.4 |
| Third midcourse correction | Service module reaction control | 241:21:59.7 | 5.7 | 2.4 | -6.48 | 36 116 | 13.795 | 173.53E | 244:22:19.1 |

**TABLE 5-IX.- ENTRY TRAJECTORY PARAMETERS**

| | |
|---|---|
| Entry interface (400 000 feet altitude) | |
| Time, hr:min:sec | 244:22:19.1 |
| Geodetic latitude, deg south ....... | 13.80 |
| Longitude, deg east | 173.52 |
| Altitude, miles | 65.8 |
| Space-fixed velocity, ft/sec ....... | 36,116 |
| Space-fixed flight-path angle, deg .... | -6.48 |
| Space-fixed heading angle, deg east of north | 98.16 |

# 6.0 LUNAR DUST

Lunar dust was evident during Apollo 12 in two respects, but in a manner which differed significantly from that observed during Apollo 11. First, the crew experienced total obscuration of visibility just prior to touchdown, and second, because of increased exposure, more dust adhered to surface equipment and contaminated the atmosphere of both spacecraft.

## 6.1 DUST EFFECTS ON LANDING VISIBILITY

During the final phase of lunar module descent, the interaction of the descent engine exhaust plume with the lunar surface resulted in the top layer of the lunar soil being eroded away. The material particles were picked up by the gas stream and transported as a dust cloud for long distances at high speeds. Crew visibility of the surface and surface features was obscured by the dust cloud.

### 6.1.1 Mechanism of Erosion

The type of erosion observed in the Apollo 11 and 12 landings is usually referred to as viscous erosion, which has been likened to the action of the wind blowing over sand dunes. The shearing force of the gas stream at the interface of the gas and lunar soil picks up the weakly cohesive particles, injects them into the stream, and accelerates the particles to high velocities. The altitude at which this erosion is first apparent and the transport rate are dependent upon the surface loading caused by the engine exhaust plume and upon the mechanical properties of the local lunar soil. This dependence is expressed in terms of several characteristic parameters, such as engine chamber pressure, exit Mach number, material density, particulate size, and cohesion. Reference 4 develops the fundamental theory for predicting erosion rates during landing and compares the analytical predictions with experimental data. A list of suitable references on this subject are contained in volume II of reference 4.

### 6.1.2 Visibility Degradation During Apollo 12

Data on the degradation of visibility during landing are derived from crew observations and photographs. The photographic record is obtained from film (fig. 6-1) exposed by a 16-mm sequence camera, which is mounted in the right-hand lunar module window. On Apollo 12 this camera was operated at 12 frames/sec. Additional photographic data on erosion are obtained from 70-mm still photographs taken in the vicinity of the lunar module during extravehicular activity. Finally, an accurate reconstruction of the trajectory from tracking and telemetry data is necessary to correlate position and time with the varying visibility conditions observed by the crew and recorded on the photographs. There is no assurance that the sequence film records the same impressions as stated by the crew for the following reasons:

a. The camera has a relatively narrow field of view compared to the crewman

b. The camera line-of-sight is more depressed toward the vertical than the crewman's normal line-of-sight; hence, the two data sources normally view different scenes

c. The range of optical response for the film is less than that of the crewman's eye

d. The environment under which the crewman made his observations is considerably different from that in which the film is viewed after the flight.

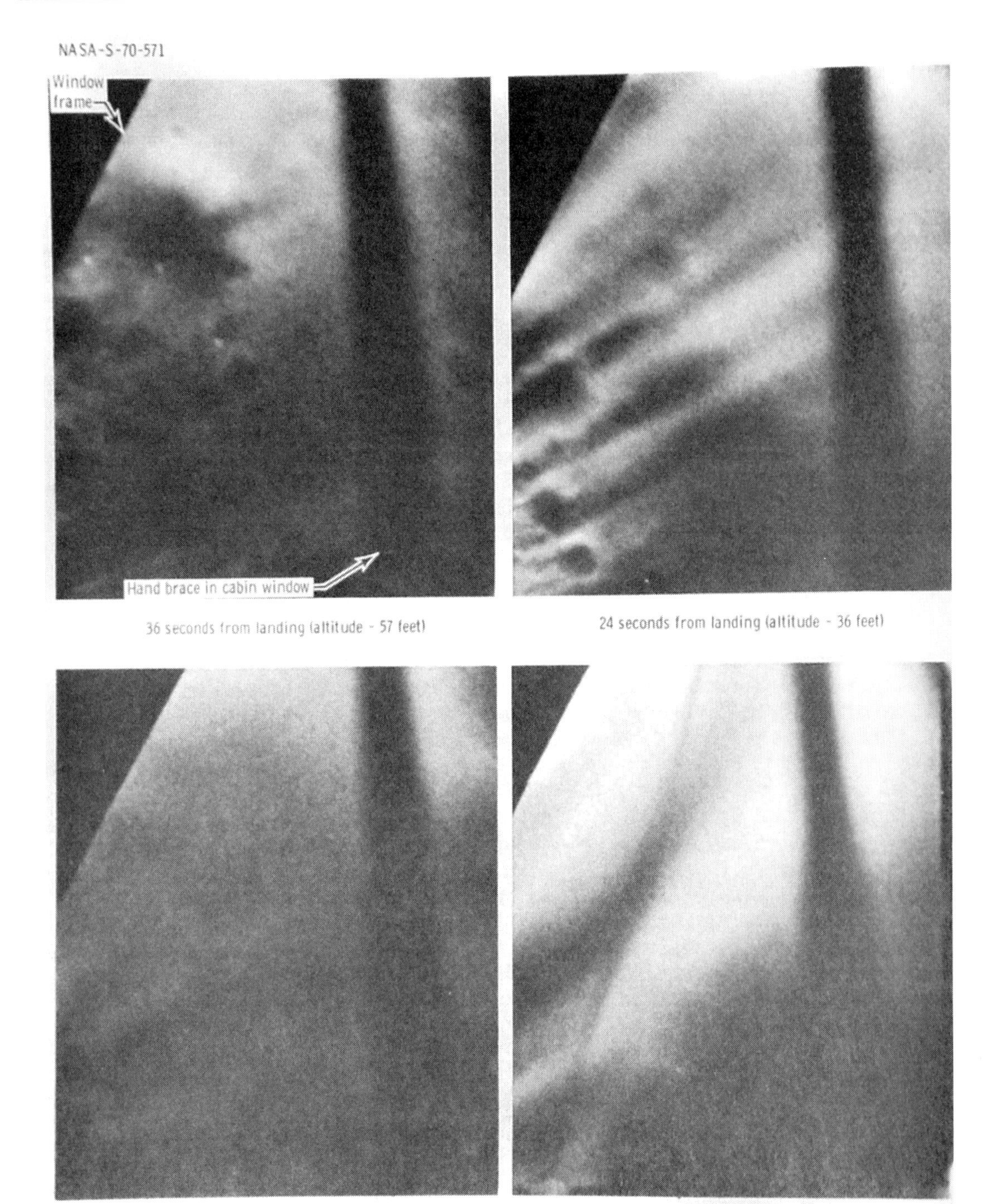


Figure 6-1 Selected sequence photographs during landing

The first time that dust is detected from the photographic observations occurs 52 seconds before touchdown. This time corresponds to an altitude of about 100 feet. There is no commentary in the voice transcription relative to dust at this point, but postflight debriefings indicate the crew noticed the movement of dust particles on the surface from a relatively higher altitude. At 180 feet altitude the Lunar Module Pilot made the comment that they could expect to get some dust before long. However, the initial effect of the dust, as first observed in the film or by the crew, indicates that there was no degradation in visibility prior to about 100 feet in altitude. However, the crew stated that dust was first observed at an altitude of about 175 feet (section 9.0). Dust continued to appear in the sequence camera photographs for the next 10 or 12 seconds as the lunar module descended to about 60 to 70 feet in altitude. Visibility is seen to have degraded, but not markedly. Beyond this point, the film shows the dust becoming more dense. Although surface features are still visible through the dust, impairment of visibility is beginning. Degradation of visibility continues until the surface is completely obscured and conditions are blind. The point at which this total obscuration occurs is somewhat subjective. At 25 seconds before touchdown, the dust cloud is quite dense, although observations of the film show some visibility of the surface. From the pilot's point of view, however, visibility is seen to be essentially zero at this time, which corresponds to an altitude of about 40 feet. Therefore, the pilot's assessment that total obscuration occurred at an altitude of about 50 feet is confirmed. The Commander considered visibility to be so completely obscured at this point that he depended entirely on his instruments for landing cues.

### 6.1.3 Comparison to Apollo 11 and Results of Analysis

Compared to the Apollo 11 landing, the degradation in visibility as a result of dust erosion was much more severe during Apollo 12. During Apollo 11, the crew likened the dust to a ground fog; that is, it reduced the visibility, but never completely obscured surface features. On Apollo 12 the landing was essentially blind for approximately the last 40 feet. In order to better understand the reasons for these differences, a detailed analysis was initiated of the factors which affect erosion and visibility. The results of that analysis, although not completed, are summarized here.

First, it was important to establish whether the surface material characteristics were different at the Apollo 11 and Apollo 12 landing sites. The various data sources provide no firm basis for a belief that a significant difference exists between the lunar material characteristics at the two sites. On the other hand, the following evidence indicates that the surface material behavior was essentially the same at the two sites

a. The height at which erosion first occurred was essentially the same on the two missions. The Apollo 11 sequence camera photographs indicate the first signs of dust at about 120 feet altitude about 65 seconds before landing.

b. Photographs taken during the extravehicular activity in the general area of the lunar module revealed that the soil disturbances caused by the descent engine exhaust produced about the same effects on the two missions.

c. Photographs of the crewmen's bootprints indicate that the soil behaved about the same at the two sites. Although there were local variations in bootprint penetrations, such variations were observed at both sites.

d. Analysis of the returned core tube samples indicates that the lunar soil had about the same density and the same particle size distribution at both sites.

Since the soil characteristics were apparently the same at the two sites, the analysis was concentrated on the aspects of the two flights that were different, that is, the descent profile over the last 200 feet of altitude and the sun elevation level at landing. Results of these analyses indicate that both of these effects contributed to the poor visibility conditions on Apollo 12. The thrust level on Apollo 12 was somewhat higher over most of the final descent and was significantly higher (about 20 percent) at about 30 feet altitude at 15 to 20 seconds before landing. This greater thrust caused a higher surface loading and therefore produced greater erosion rates. More significant, however, was the effect of the lower sun angle (5.1 degrees on Apollo 12 compared to 10.8 degrees on Apollo 11). For given dust cloud density the combined effects of light attenuation, veiling luminance, and a diffuse illumination on the surface are much more serious at the lower sun angle and can be shown analytically to produce the effects observed on Apollo 12. Analysis is continuing on a parametric variation of the factors which affect erosion and visibility. However, all these analyses are based upon certain assumptions about the optical scattering properties of the lunar dust and upon an idealized lunar model. Thus, these limitations make it impossible to conclusively prove that the effects noted can indeed be attributed to the sum elevation angle. Undeterminable differences in critical soil properties, such as cohesion, could have produced the same effects.

### 6.1.4 Instrument Landing Procedures

Preliminary studies show the impracticality of various means for reducing the dust effects on visibility, largely because of the weight and performance limitations of the spacecraft. The lunar module was designed with the capability to be flown entirely on instruments during the landing phase. The two accomplished lunar landings have provided the confidence that an instrument landing is within the capability of the spacecraft systems. Therefore, on Apollo 13, onboard software will be modified to permit reentry into an automatic descent program after manual modes have been exercised. This change will allow selection or redesignation of a suitable landing site, followed by automatic nulling of horizontal rates and automatic vertical descent from the resulting hover condition, which would occur at an altitude above appreciable dust effects.

### 6.2 CONTAMINATION OF THE SPACECRAFT ATMOSPHERE

The amount of lunar dust encountered by the Apollo 12 crew appeared to be appreciably greater than in Apollo 11. This condition manifested itself by contaminating the atmospheres in both spacecraft and depositing dust over much of the lunar surface equipment and onboard systems. The cohesive properties of lunar dust in a vacuum, augmented by electrostatic properties, tend to make it adhere to anything it contacts. These properties diminish in the presence of the gas of an atmosphere. Upon attaining zero gravity, some of the lunar dust floats up in the cabin atmosphere and becomes widely dispersed. This process tends to be continuous, and renders present atmosphere filtration techniques inadequate. The presence of the lunar dust in the cabin of either spacecraft does not detrimentally affect the operation of onboard systems, but the dust could present a hazard to crew health, and at least it constitutes a nuisance. The potential health hazards are eye and lung contamination when the dust floats in zero g. In an effort to minimize this nuisance on future flights, various dust removal techniques were evaluated for cleaning the spacesuits and equipment on the lunar surface prior to ingressing the lunar module.

## 7.0 COMMAND AND SERVICE MODULE PERFORMANCE

Performance of command and service module systems is discussed in this section. The sequential, pyrotechnic, earth landing, and emergency detection systems operated as intended and are not discussed further. Discrepancies and anomalies in command and service module systems are generally mentioned in this section but are discussed in greater detail in the anomaly summary section 14.1.

### 7.1 STRUCTURAL AND MECHANICAL SYSTEMS

At earth lift-off, measured winds, both at the surface and in the region of maximum dynamic pressure, indicate that structural loads were well below the established limits. The predicted and calculated spacecraft loads at lift-off, in the region of maximum dynamic pressure, at the end of first stage boost, and during staging were similar to or less than for Apollo 11. Command module accelerometer data prior to S-IC center-engine cutoff indicate a sustained 5-hertz longitudinal oscillation of 0.2g amplitude, which is similar to that measured during Apollo 4. The vibration reported by the crew during the S-II boost phase had a measured amplitude of less than 0.058 at a frequency of 15 hertz. However, the amplitudes of both oscillations were within acceptable spacecraft structural design limits. All structural loads during S-IVB boost, translunar injection, both docking operations, all service propulsion maneuvers, and entry were also well within design limits.

As with all other mechanical systems, the docking system performed as required for both the translunar and lunar orbit docking events and sustained contact conditions consistent with those during Apollo 9, 10, and 11.

The temperatures of all passively controlled elements remained within acceptable limits. However during transearth flight, a temperature transducer, located on the service propulsion system fuel storage tank, exhibited a temperature increase approximately twice the rate observed on previous missions. This anomaly is discussed further in section 7.5. Five thermal transducers on the service module failed as a result of a potential electrical discharge at 36.5 seconds after lift-off. These measurements were not critical to crew safety, and the loss did not constitute a problem. This anomaly is also discussed in sections 7.5 and 14.1.3.

The lunar module crew reported seeing a piece of strap-like material in the vicinity of the service module/adapter interface just prior to docking (discussed in section 14.1.8). The crew also reported streaks on the command module windows after translunar injection, as discussed in section 14.1.11. In addition, an oxygen hose retention bracket became unbonded from its support bracket at earth landing (as discussed in section 14.1.14), and a piece of lanyard for the forward heat shield was missing during postflight inspection (as discussed in section 14.1.16).

## 7.2 ELECTRICAL POWER

### 7.2.1 Power Distribution

The electrical power distribution and sequential systems performed satisfactorily throughout the flight. At 36.5 seconds into the flight, the spacecraft was subjected to a potential discharge between space vehicle and ground. A voltage transient, induced on the battery relay bus by the static discharge, tripped the silicon controlled rectifiers in the fuel cell overload sensors and disconnected the fuel cells from the bus. As a result, the total main bus load of 75 amperes was being supplied by entry batteries A and B. The main bus voltage dropped momentarily to 18 or 19 volts but recovered to 23 or 24 volts within a few milliseconds. The low voltage on the main dc buses caused the undervoltage warning lights to illuminate, the signal conditioning equipment to drop out, and the input to the inverter to decrease momentarily. The momentary low-voltage to the inverters resulted in a low output ac voltage, which tripped the ac undervolt age sensor and caused the ac bus 1 fail light to illuminate. The transient that tripped the fuel cell overload circuitry also tripped the inverter overload circuitry, thereby causing the ac overload lights to illuminate. See section 14.1.3 for a more complete discussion of the potential electrical discharge events.

The crew checked the ac and dc buses on the selectable meter and ascertained that the electrical power system was still functional. At 00:02:22, fuel cell power was restored to the buses, and bus voltage remained normal for the remainder of the flight. During earth-orbital insertion checks, a circuit breaker was found in an open position and is discussed further in section 14.1.4.

### 7.2.2 Fuel Cells

The fuel cells were activated 64 hours prior to launch, conditioned for 6-1/2 hours, and then placed on open-circuit inline heater operation until cryogenic loading was completed. After loading, fuel cell 2 was placed on the line and supplied a current of about 20 amperes as part of the prelaunch cryogenics management plan. All three fuel cells were placed on the bus 3-1/2 hours prior to launch. Differences in initial load sharing between fuel cells were as great as 9 amperes because of launch cryogenic management requirements. The load sharing gradually stabilized to a maximum deviation of 2 or 3 amperes early in the flight.

During the mission, the fuel cells supplied approximately 501 kW-h of energy at an average current of 23.2 amperes per fuel cell and an average bus voltage of 29.4 volts.

All fuel cell thermal parameters remained within normal operating limits and agreed with predicted flight values. However, the condenser exit temperature on fuel cell 2 fluctuated periodically every 3 to 8 minutes throughout the flight. This disturbance was similar to that observed on all other flights and is discussed in more detail in reference 8. The periodic disturbance has been shown to have no effect on fuel cell performance.

The regulated hydrogen pressure of fuel cell 3 appeared to decrease slowly by about 2 psi during the mission. The apparent cause of the decay was a drift in the output of the pressure transducer (as discussed in section 14.1.17) that resulted from hydrogen leaking into the evacuated reference cavity of the transducer.

### 7.2.3 Batteries

At 36.5 seconds, when the fuel cells disconnected from the bus, entry batteries A and B assumed the total spacecraft load. Entry battery C is intentionally isolated during the flight until entry to maximize crew safety. This step increase in current from approximately 4 amperes to 40 amperes on each of the batteries (A and B) resulted in a low-voltage transient. However, within approximately 134 milliseconds of the fuel cell disconnection, the logic bus voltage data showed the battery bus voltage had increased to 25.2 V dc. The battery bus voltage had increased to 26 V dc at the time the fuel cells were placed back on the main buses.

Entry batteries A and B were both charged once at the launch site and six times during flight with nominal charging performance. Load sharing and voltage delivery were satisfactory during each of the service propulsion firings. The batteries were essentially fully charged at entry and performance was nominal.

## 7.3 COMMUNICATIONS EQUIPMENT

The communications system satisfactorily supported the mission except for the following described conditions. Uplink and downlink signal strengths were, on a number of occasions, below expected levels for normal high-gain antenna performance, which is discussed further in section 14.1.6. VHF voice communications between the command module and the lunar module were unacceptable during the ascent, rendezvous, and docking portions of the mission. Section 14.1.19 contains a detailed discussion of this problem. The S-band communications system provided excellent quality voice throughout the mission, as did the VHF/AM system during the earth-orbital and recovery portions of the mission. The spacecraft omnidirectional antenna system was used for communications during most of translunar and transearth coast. During operation on these antennas, the maximum level of received carrier power agreed with predictions.

Two ground-plane radials associated with VHF recovery antenna 2 did not deploy properly. However, VHF voice communications with recovery forces were not affected, and further details concerning this problem are presented in section 14.1.12.

## 7.4 CRYOGENIC STORAGE

During cryogenic loading approximately 51 hours before the scheduled launch, the performance of hydrogen tank 2 was unacceptable in that the tank filled much slower than normal and had a high boiloff rate during the stabilization period. A visual inspection of the tank revealed a thick layer of frost on the tank exterior, indicating loss of the vacuum in the insulating annulus. The tank was replaced with a tank from the Apollo 13 spacecraft, and cryogenic loading was satisfactorily completed. A detailed discussion of the hydrogen tank malfunction is provided in section 14.1.2.

Cryogenics were satisfactorily supplied to the fuel cells and to the environmental control system throughout the mission. At launch, 635 pounds of oxygen and 53.8 pounds of hydrogen were available, and at command module/service module separation, 150 pounds of oxygen and 9.6 pounds of hydrogen remained. The predicted oxygen and hydrogen quantities remaining at command module/service module separation were 155 pounds and 8.2 pounds, respectively. The rate of oxygen depletion was higher than the expected values by approximately 0.1 pound per hour. A detailed discussion of this problem is provided in section 14.1.7. Hydrogen consumption was normal during the flight.

## 7.5 INSTRUMENTATION

As a result of the potential electrical discharge at 36.5 seconds after lift-off, five temperature measurements and four pressure/temperature measurements failed. These measurements were all located in the same general plane of the service module. Analysis of the temperature sensor failures indicates the most probable cause to be an electrical overstress of a diode or resistor in a measurement zone box. Failure of the pressure/ temperature measurements apparently was caused by an electrical overstress of the semiconductor strain gages, located on the pressure-sensing diaphragm, or of the bridge voltage-regulating Zener diode. A detailed discussion of this anomaly is presented in section 14.1.3.

The central timing equipment and the signal conditioning equipment also were temporarily affected by the potential discharges at 36.5 and 52 seconds. The time reference in the central timing equipment jumped ahead at 36.5 seconds and was erratic until 52.49 seconds, when it reset to zero. The central timing equipment performed satisfactorily thereafter. The signal conditioning equipment was turned off by its undervoltage sensor at 36.5 seconds, when the bus voltage dropped below 22.9 V dc. The signal conditioning equipment returned to operation at 97 seconds, when the bus voltage had recovered to normal levels.

During the flight, several other problems were noted. During the first 30 hours, the reaction control quad D helium manifold pressure drifted high by approximately 14 psi. At 160:07:00, the measurement dropped to a reading of 30 psi low. The problem involves two independent failures and is discussed in section 14.1.17.

The temperature sensor for the service propulsion fuel storage tank failed during preflight testing at the launch site, and the sensor/signal conditioner system was replaced. The response of this temperature measurement during the flight was greater than anticipated. While the original sensor was located under the tank insulation, a postflight investigation has established that the replacement sensor was located on an uninsulated portion of the tank. At this location, the high temperature-response rate would be expected.

During most of the mission, the suit pressure transducer indicated 0.4 to 0.5 psi lower than cabin pressure and, at one time, indicated as low as 0.1 psia. This anomaly is discussed in section 14.1.17.

The carbon dioxide sensor did not function during the mission. This type of sensor has a history of erratic operation, and previous testing has shown it to be sensitive to moisture contamination.

The primary water/glycol pump outlet pressure was indicating from 3.5 to 4.5 psi higher than normal prior to launch and throughout the flight. A similar calibration shift has occurred previously and has typically resulted from inadvertent system overpressurization. A detailed review of data derived since the last transducer calibration by the contractor revealed only one minor overpressurization, which had no apparent effect on the transducer. However, such an occurrence is still considered the most probable cause of the discrepancy.

The potable water quantity transducer operated erratically prior to launch and during the flight. Although similar anomalous operation occurred during Apollo 8 as a result of moisture contamination, testing after Apollo 12 revealed a film contamination on the extreme surfaces of the resistance wafer. Section 14.1.17 has additional discussion of this malfunction.

The regulated hydrogen pressure for fuel cell 3 gradually decayed during the flight. Fuel cell performance was satisfactory, and the pressure decrease was attributed to failure of the pressure transducer. The probable failure mode is a hydrogen leak around the transducer diaphragm into the vacuum reference chamber, thus decreasing the normal differential pressure across the diaphragm. Similar transducer failures have occurred during fuel cell ground tests.

## 7.6 GUIDANCE, NAVIGATION, AND CONTROL

Command module guidance, navigation, and control system performance was satisfactory throughout the mission. Because of the static discharges experienced during earth ascent and described in detail in section 14.1.3, the normal ascent monitoring functions were not performed. As a result of one of these discharges, the inertial reference was lost and the inertial platform was subsequently powered down; therefore, it became necessary to perform both an orientation determination (computer program P51) and a platform alignment (P52) in earth orbit. In addition, an extra platform alignment on the second night pass was conducted to detect any detrimental effects of the static discharge on inertial component performance. As shown in table 7.6-I, the gyro performance determined from these and all subsequent alignments during the mission was excellent.

System monitoring of translunar injection and control during transposition and docking were normal, although the entry-monitor-system velocity counter did not reflect the velocity changes expected by the crew during transposition. The apparent discrepancies were caused by an acceptable accelerometer bias of 0.023 ft/sec$^2$. This bias remained essentially constant throughout the mission and is shown in table 7.6-II, which contains entry monitor system parameters for each service propulsion system maneuver.

### TABLE 7.6-I.- PLATFORM ALIGNMENT SUMMARY

| Time, hr:min | Program option* | Star used | Gyro torquing angle, deg | | | Star angle difference, deg | Gyro drift, mERU | | | Comments |
|---|---|---|---|---|---|---|---|---|---|---|
| | | | X | Y | Z | | X | Y | Z | |
| 00:24 | | | | | | | | | | |
| 00:52 | | | | | | 0.01 | | | | Program 51 |
| 00:52 | 1 | 14 Canopus, 15 Sirius | +0.755 | +0.941 | -0.366 | 0.01 | — | — | — | |
| 02:20 | 3 | 01 Alpheratz, 45 Fomalhaut | -0.014 | -0.028 | +0.018 | 0.00 | +0.8 | +1.7 | +1.1 | |
| 05:53 | 1 | 14 Canopus, 16 Procyon | +0.764 | +0.576 | -1.187 | 0.01 | — | — | — | |
| 14:57 | 3 | 16 Procyon, 12 Rigel | +0.127 | -0.171 | -0.281 | 0.00 | -0.9 | +1.3 | -2.1 | |
| 29:48 | 3 | 24 Gienah, 27 Alkaid | +0.250 | -0.246 | +0.125 | 0.01 | -1.1 | +1.1 | +0.6 | Check star 22 Regulus |
| 55:02 | 3 | 03 Navi, 13 Capella | +0.515 | -0.492 | +0.289 | 0.01 | -1.4 | +1.3 | +0.8 | Check star 20 Dnoces |
| 78:21 | 3 | 03 Navi, 13 Capella | +0.400 | -0.462 | +0.263 | 0.02 | -1.1 | +1.3 | +0.8 | |
| 81:06 | 1 | 01 Alpheratz, 10 Mirfak | +0.180 | +0.259 | +0.658 | 0.02 | — | — | — | |
| 86:45 | 3 | 7 Menkar, 13 Capella | +0.078 | -0.111 | +0.090 | 0.02 | -0.9 | +1.3 | +1.05 | Check star 11 Aldebaran |
| 88:55 | 3 | 16 Procyon, 20 Dnoces | +0.013 | -0.029 | +0.069 | 0.02 | -0.4 | +0.9 | +2.1 | Check star 22 Regulus |
| 102:50 | 1 | 20 Dnoces, 27 Alkaid | +0.238 | -0.294 | +0.175 | 0.01 | — | — | — | |
| 108:49 | 3 | 11 Aldebaran, 10 Mirfak | +0.135 | -0.061 | +0.000 | 0.01 | -1.5 | +0.7 | 0.0 | |
| 110:44 | 3 | 21 Alphard, 26 Spica | -0.035 | -0.056 | +0.44 | 0.01 | +1.2 | +1.9 | +1.5 | |
| 118:32 | 1 | 03 Navi, 20 Dnoces | +0.562 | +0.000 | +0.670 | 0.02 | — | — | — | Check star 13 Capella |
| 120:35 | 1 | 12 Rigel, 21 Alphard | -0.708 | -0.961 | -0.392 | 0.02 | — | — | — | |

| | | | | | | | | | | |
|---|---|---|---|---|---|---|---|---|---|---|
| 132:45 | 3 | 12 Rigel, 21 Alphard | +0.255 | -0.228 | +0.141 | — | -1.4 | +1.3 | +0.8 | |
| 138:20 | 3 | 22 Regulus, 26 Spica | +0.088 | -0.160 | +0.102 | 0.02 | -1.0 | +1.8 | +1.2 | |
| 140:17 | 3 | 11 Aldebaran, 20 Dnoces | +0.022 | -0.021 | -0.043 | 0.01 | -0.8 | +0.7 | -1.5 | |
| 142:19 | 3 | 23 Denebola, 26 Spica | +0.028 | -0.044 | +0.019 | 0.00 | -0.9 | +1.5 | +0.6 | |
| 158:17 | 1 | 22 Regulus, 27 Alkaid | -0.382 | -0.048 | +0.331 | — | — | — | — | |
| 159:16 | 1 | | -84.79 | | -49.479 | — | — | — | — | Pulse torqued to orient |
| 159:54 | 3 | 16 Procyon, 23 Denebola | +0.065 | -0.037 | -0.098 | 0.01 | | | | |
| 164:06 | 3 | 21 Alphard, 26 Spica | +0.095 | -0.088 | -0.003 | 0.03 | -1.5 | +1.4 | -0.1 | |
| 165:52 | 3 | 20 enoces, 21 Alphard | +0.023 | -0.003 | +0.073 | — | — | — | — | |
| 167:57 | 3 | 16 Procyon, 20 Dnoces | +0.053 | -0.032 | +0.003 | 0.02 | -1.7 | +1.0 | +0.1 | |
| 173:33 | 1 | | | | | | | | | Pulse torqued to orient |
| 173:52 | 3 | 04 Achernar, 22 Regulus | -0.216 | -0.004 | +0.149 | 0.01 | — | — | — | |
| 187:55 | 3 | 06 Acamar, 45 Fomalhaut | +0.288 | -0.301 | +0.211 | 0.01 | -1.4 | +1.4 | +1.0 | |
| 210:09 | 3 | 34 Atria, 30 Menkent | +0.414 | -0.396 | +0.240 | 0.02 | -1.2 | +1.2 | +0.7 | Check star 25 Acrux |
| 211:30 | 3 | 15 Sirius, 12 Rigel | +0.034 | -0.053 | -0.002 | 0.04 | — | — | — | No torque |
| 211:37 | 3 | 45 Fomalhaut, 34 Atria | +0.061 | -0.009 | -0.004 | 0.02 | — | — | — | No torque |
| 221:39 | 3 | 25 Acrux, 17 Regor | +0.191 | -0.199 | +0.149 | — | -1.1 | +1.2 | +0.8 | |
| 240:08 | 1 | 35 Rasalhague, 41 Dabih | +0.195 | -0.544 | -0.641 | 0.00 | — | — | — | Check star 37 Nunki |
| 243:01 | 3 | 23 Denebola, 30 Menkent | +0.053 | -0.069 | +0.015 | 0.01 | -1.2 | +1.6 | +0.4 | |

*1 - Preferred; 2 - Nominal; 3 - REFSMMAT; 4 - Landing site.

## TABLE 7.6-II.- ENTRY MONITOR SYSTEM PERFORMANCE

| | Maneuver | | | | | | | |
|---|---|---|---|---|---|---|---|---|
| | First midcourse correction | Lunar orbit insertion | Circular-ization | First plane change | Second plane change | Trans-earth injection | Second midcourse correction | System test[a] |
| Total velocity change, ft/sec | +61.7 | +2889.3 | +165.5 | +349.7 | +381.3 | +3042.3 | +2.0 | |
| Velocity change set into counter, ft/sec | +57.2 | +2882.4 | +159.4 | +337.1 | +368.2 | +3021.1 | +2.0 | 0 |
| Estimated time of counter operation sec | 39 | 388 | 47 | 48 | 49 | 160 | 38 | 100 |
| Planned residual, ft/sec | -4.2 | +1.0 | -4.4 | -8.4 | -11.3 | -14.4 | +1.8 | n/a |
| Actual counter residual, ft/sec[b] (corrected) | -4.4 | -6.8 | -5.6 | -12.6 | -13.5 | -21.0 | +0.2 | n/a |
| Entry monitor system error, ft/sec | -0.2 | -7.8 | -1.2 | -4.2 | -2.2 | -6.6 | -1.6 | -2.2 |
| Estimated bias[c], ft/sec/sec | -.005 | -.020 | -.025 | -.055 | -.045 | -.045 | -.042 | -0.022 |

[a] Performed at 238 hours.

[b] A value of 0.2 ft/sec and the observed command module computer X-axis residual were added to determine the corrected error.

[c] Corrected error divided by estimated counter operating time, i.e. firing time plus 30 seconds.

Table 7.6-III contains a summary of selected guidance and control parameters for executed maneuvers. All maneuvers were nominal, although the crew reported a "dutch roll" sensation during the second plane change maneuver in lunar orbit. Figure 7.6-1 contains a time history of selected control parameters for a portion of that maneuver and a similar set of parameters for a like portion of the transearth injection maneuver. The spacecraft response during both maneuvers is comparable to that noted on previous missions and within the range of responses expected under randomly initiated fuel slosh.

All attitude control functions throughout the mission were normal, with passive thermal control again proving to be an excellent method for conserving propellant during translunar and transearth coast. Two pairs of reaction control engines fired for an abnormally long time during the initial sleep period in lunar orbit. The docked spacecraft were in attitude hold with a 10-degree deadband to provide thermal control. Because of gravity-gradient torques, the digital autopilot was expected to maintain attitudes near one edge of the deadband using minimum-impulse firings of 14 milliseconds duration. However, the data show that one pair of engines (pitch) fired for 440 milliseconds and another pair (yaw) fired for 755 milliseconds, with all four engines commanded on simultaneously. A detailed analysis indicates the most likely cause of these long firings was a transient in an electronic coupling display unit. Because of the orientation of the inertial platform to the spacecraft, a transient of 0.38 degree about the platform Y gimbal axis would cause attitude errors of minus 0.23 degree and minus 0.30 degree about the pitch and yaw body axes, respectively. The calculated firings times required to correct for these attitude errors and their associated rates agree well with the observed firing times. Ground tests have demonstrated that in the coupling display unit, transients are caused by the charging and discharging of capacitors associated with certain transistorized switch circuits. The transients are especially noticeable when certain switches are energized after a long period of inactivity especially when several switch circuits experience such a state change simultaneously. Analysis of these transients and the related thruster firing combinations will continue, with results to be presented in a supplemental report (appendix E).

The Command Module Pilot reported that the coelliptic sequence initiation solution in the command module computer did not converge to match those from the ground and the lunar module until a large number of VHF ranging and optical marks had been taken. Analysis indicates that the initial VHF ranging input was incorrect and degraded the onboard state vector. The source of the incorrect VHF input is not known; however, there is a discrepancy in the computer interface logic which can cause the range to be read out incorrectly. Under certain low-probability conditions, one or more of the synchronizing pulses, with which the computer shifts the digital range word out of the VHF, can be split and recognized as two pulses. The magnitude of the resulting range error is dependent on the significance of the affected bit. The computer program protects against an erroneous input by inhibiting automatic state vector updates larger than a preset threshold (2000 feet or 2 feet per second). If an update is larger than this threshold, it is displayed to the crew for manual acceptance or rejection. Updates are normally rejected if provisionally displayed except at the beginning of a sequence of marks when the state vector can be expected to be degraded, as was the case for the first UHF mark.

## TABLE 7.6-III.- GUIDANCE AND CONTROL MANEUVER SUMMARY

| | Maneuver | | | | | |
|---|---|---|---|---|---|---|
| Parameter | First midcourse correction | Lunar orbit insertion | Lunar orbit circularization | First plane change | Second plane change | Transearth injection |
| Time | | | | | | |
| Ignition, hr:min:sec | 30:52:44.36 | 83:25:23.36 | 87:48:48.08 | 119:47:13.23 | 159:04:45.47 | 172:27:16.81 |
| Cutoff, hr:min:sec | 30:52:53.55 | 83:31:15.61 | 87:49:04.99 | 119:47:31.46 | 159:05:04.72 | 172:29:27.13 |
| Duration, min:sec | 0:09.19 | 5:52.25 | 0:16.91 | 0:18.23 | 0:19.25 | 2:10.32 |
| Velocity gained, ft/sec* (desired/actual) | | | | | | |
| X | +19.60/+19.70 | -1401.93/-1401.93 | -159.86/-159.59 | +44.05/+44.11 | +23.23/+23.06 | -1772.09/-1771.92 |
| Y | +41.10/+41.60 | -1224.43/-1224.74 | -13.60/-13.70 | +197.26/+197.72 | +214.51/+215.06 | +2244.9./+2245.22 |
| Z | -41.61/-42.54 | -2209.88/-2210.05 | -40.59/-40.55 | -285.36/-285.27 | -314.30/-314.31 | +1036.97/+1036.24 |
| Velocity residual, ft/sec (spacecraft coordinates)** | | | | | | |
| X | -0.1 | -0.2 | +0.3 | -0.3 | -0.7 | -0.1 |
| Y | -0.3 | 0.0 | 0.0 | +0.1 | +0.3 | +0.6 |
| Z | 0.0 | +0.1 | +0.1 | +0.4 | +0.6 | +0.1 |
| Entry monitor system | -0.2 | -7.8 | -1.2 | -4.2 | -2.2 | -6.6 |
| Engine gimbal position, deg | | | | | | |
| Initial | | | | | | |
| Pitch | +0.99 | +0.94 | +1.51 | -0.65 | -0.70 | -0.57 |
| Yaw | -0.18 | -0.10 | -0.54 | +0.54 | +0.33 | +0.28 |
| Maximum excursion | | | | | | |
| Pitch | +0.39 | +0.35 | +0.31 | -1.98 | -2.10 | -2.06 |
| Yaw | -0.38 | -0.34 | -0.24 | +1.53 | +2.04 | +1.78 |
| Steady-state | | | | | | |
| Pitch | +1.21 | +1.08 | +1.78 | -0.31 | -0.18 | -0.31 |
| Yaw | +0.20 | +0.07 | -0.35 | +0.71 | +0.75 | +0.45 |
| Cutoff | | | | | | |
| Pitch | +1.21 | +1.68 | +1.58 | -0.44 | -0.35 | -0.48 |
| Yaw | -0.01 | -0.31 | -0.42 | +0.54 | +0.45 | -1.20 |

| | | | | | | |
|---|---|---|---|---|---|---|
| Maximum rate excursion, deg/sec | | | | | | |
| Pitch | +0.04 | -0.04 | -0.04 | +1.27 | +1.67 | +1.39 |
| Yaw | +0.08 | +0.12 | +0.20 | -0.60 | -0.68 | -0.51 |
| Roll | -0.04 | -0.04 | -0.45 | -0.85 | +1.01 | -0.89 |
| Maximum attitude error, deg | | | | | | |
| Pitch | -0.08 | +0.19 | +0.24 | +0.08 | +0.37 | -0.24 |
| Yaw | +0.20 | -0.08 | -0.10 | -0.28 | +0.32 | -0.28 |
| Roll | -0.13 | -5.00*** | -2.40 | -4.28 | +0.42 | -5.00 |

*Velocity residuals in spacecraft coordinates after trimming has been completed.
**Velocity gained in earth- or moon-centered inertial coordinates.
***Telemetry signal saturated.

VHF and optics marks following this initial input resulted in consistently large corrections until after ten optics and fourteen VHF updates had been incorporated. Thereafter, state vector updates became smaller, and the second attempt to obtain a solution indicated close agreement with the two independent solutions. No further difficulty was encountered throughout the rendezvous sequence, although the loss of the tracking light after coelliptic sequence initiation precluded the taking of optics marks during darkness.

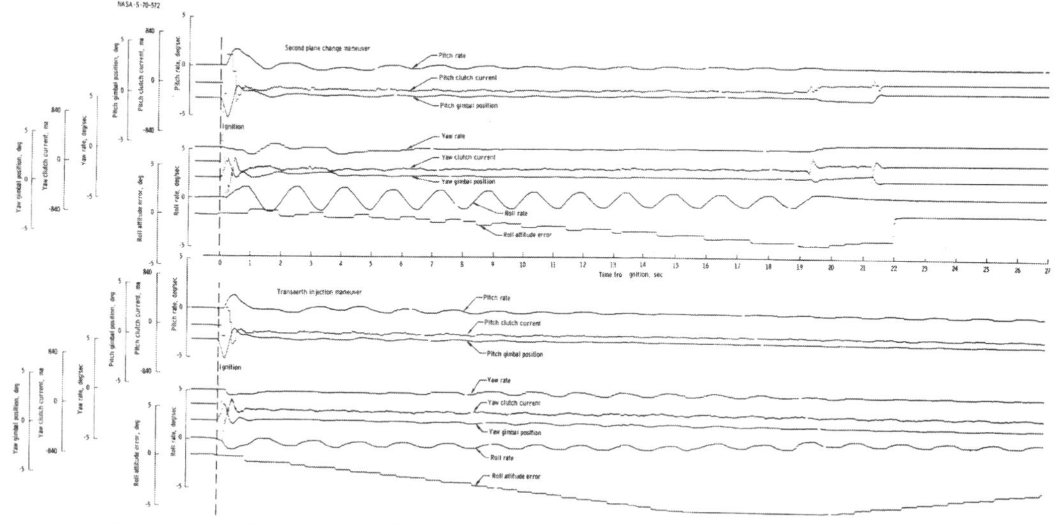


Figure 7.6-1 Comparison of spacecraft dynamics during second plane change and transearth injection maneuvers.

Midcourse navigation using star horizon measurements was performed during translunar and transearth coast as in previous lunar missions. The transearth measurements, however, were taken in an attempt to establish the effect on visual observations of sun incidence at various angles to the line of sight. Preliminary indications are that the desired data were obtained.

A number of orbit navigation exercises using landmark tracking techniques were conducted in lunar orbit. No difficulties were experienced.

Entry was performed under automatic control as planned. Spacecraft response was normal and similar to that seen on previous missions. Earth landing occurred approximately 1.1 miles from the target.

The preflight and inflight performance history of the inertial components is summarized in table 7.6-IV. As shown, the deviations in those error sources measurable in flight indicate excellent component performance. Because of the loss of platform reference during launch (discussed in section 14.1.3), no ascent velocity comparisons with the S-IVB platform could be made.

The computer performed as intended throughout the mission. A number of alarms occurred, but each is explainable by either a procedural error or by the two static discharges.

Approximately 1-1/2 hours before launch, the crew noted an all-"8's" indication on the main display and keyboard assembly. As experienced in several ground tests, contamination in certain relays can cause this discrepant indication. Section 14.1.1 contains a more detailed discussion of this problem.

**TABLE 7.6-IV.- INERTIAL COMPONENT PREFLIGHT HISTORY - COMMAND MODULE**

| Error | Sample mean | Standard deviation | No. of samples | Countdown value | Flight load | Inflight performance |
|---|---|---|---|---|---|---|
| | | Accelerometers | | | | |
| X - Scale factor error, ppm ..... | -173 | 40 | 7 | -202 | -220 | — |
| Bias, $cm/sec^2$ .......... | -0.01 | 0.13 | 7 | -0.09 | -0.09 | 0.0 |
| Y - Scale factor error, ppm ..... | -243 | 65 | 9 | -330 | -350 | — |
| Bias, $cm/sec^2$ .......... | -0.13 | 0.05 | 9 | -0.08 | -0.09 | -0.15 |
| Z - Scale factor error, ppm ..... | -306 | 59 | 7 | -419 | -370 | — |
| Bias, $cm/sec^2$ .......... | -0.19 | 0.03 | 7 | -0.13 | -0.16 | -0.16 |
| | | Gyroscopes | | | | |
| X - Null bias drift, mERU ...... | -1.5 | 1.8 | 9 | -1.3 | -0.1 | -0.9 |
| Acceleration drift, spin reference axis, mERU/g ......... | -1.4 | 5.3 | 7 | -3.5 | -4.0 | — |
| Acceleration drift, input axis, mERU/g ......... | 6.7 | 6.7 | 7 | 18.2 | 13.0 | — |
| Y - Null bias drift, mERU ...... | -0.6 | 0.8 | 9 | 0.2 | -0.1 | 1.3 |
| Acceleration drift, spin reference axis, mERU/g ......... | -3.3 | 0.4 | 7 | -3.3 | -4.0 | — |
| Acceleration drift, input axis, mERU/g ......... | 0.7 | 0 | 7 | 1.7 | 0.0 | — |
| Z - Null bias drift, mERU ...... | -2.8 | 1.3 | 9 | -1.4 | 40.1 | +0.5 |
| Acceleration drift, spin reference axis, mERU/g ......... | -3.5 | 4.2 | 7 | -4.6 | -6.0 | — |
| Acceleration drift, input axis, mERU/g ......... | -0.1 | 2.3 | 7 | 0.1 | -1.0 | — |

The sextant and the scanning telescope performed normally with the exception of a random shaft axis movement noted when the system was operated in the zero-optics mode. See section 14.1.9 for details.

The stabilization and control system performed properly throughout the mission. Several gyro display coupler drift checks were obtained during transearth high-gain antenna tests. The relatively large drift values evident in the first test, as indicated in the following table, were caused by the large yaw angle to which the system was aligned, since degradation in drift as yaw angle increases is normal for this type of mechanization.

| Time | Body-mounted attitude gyro Package | Measured drift rate, deg/hr | | |
|---|---|---|---|---|
| | | Roll | Pitch | Yaw |
| 193:58 | 2 | 24.0 | 15.1 | 5.5 |
| 214:43 | 1 | 4.5 | 4.4 | 3.6 |
| 216:33 | 1 | 3.2 | 3.7 | 3.4 |
| 218:16 | 2 | 1.8 | 4.1 | 4.8 |

## 7.7 REACTION CONTROL

### 7.7.1 Service Module

The usable propellant loaded was 1341 pounds, of which 961 pounds, approximately 275 pounds more than predicted, were consumed. Propellant utilization was near that predicted through spacecraft/S-IVB separation. After separation and through the beginning of the first passive thermal control period, all digital autopilot maneuvering was performed using a 0.5 deg/sec maneuver rate, instead of the 0.2 deg/sec rate used for propellant usage predictions. Therefore, about 90 pounds more propellant were used during this period than expected. Propellant usage from this time to rendezvous was near predictions. Again, during lunar orbit photography, more propellant was used than was predicted. Quad package temperatures were satisfactorily maintained between 119° and 145° F, except after periods of high engine activity where a maximum temperature of 170° F was noted. System pressures were also maintained within regulated limits, indicating proper component performance.

The backup onboard and telemetry instrumentation for propellant gaging on all quads was lost at 36.5 seconds after lift-off (discussed in section 14.1.3). The quad D helium manifold pressure transducer also malfunctioned during the mission. Unreal and erratic readings from 194 to 148 psia were experienced throughout the mission. However, the quad D fuel and oxidizer pressure transducers provided adequate data to insure that the system was operating normally.

The crew reported that one helium and one propellant isolation valve inadvertently went to the closed position at the time of pyrotechnic separation of the command and service modules from the S-IVB. Inadvertent valve closures were also noted at separation during Apollo 9 and 11. The valves were reopened in accordance with a standard procedure and operated properly thereafter.

### 7.7.2 Command Module

System pressures and temperatures from launch to activation were stable. Helium tank temperatures varied between 54° and 75° F throughout the mission. System activation and checkout were normal. The helium source pressures stabilized at 3540 psia after activation, and the regulated pressures stabilized at 292 psia. Propellant consumption from system 1, which was used during entry, was 35 pounds and all parameters were normal.

During postflight decontamination procedures, the system 1 oxidizer isolation valve would stay in the open but not the closed position. The valve, however, did reposition to the open and closed positions properly when commanded. Section 14.1.13 contains a detailed discussion of this problem. During postflight testing, the two wires to the automatic coil of the fuel valve of the minus roll engine (no. 4) in system 2 were found to be severed. Because the break shows no salt-water corrosion, which would be expected if the severing occurred before spacecraft retrieval, it is concluded the wires were inadvertently broken during postflight handling. Therefore, the wire failure could not have affected flight performance, had system 2 been required for entry.

## 7.8 SERVICE PROPULSION SYSTEM

Service propulsion system performance was satisfactory during each of the six maneuvers, as indicated by steady-state pressure and gaging system data and the actual velocity gained. The system had a total firing time of approximately 547 seconds. The ignition times and firing durations are contained in table 7.6-III. The longest engine firing was the 352.2-second lunar orbit insertion maneuver. The third, fourth, fifth, and sixth service propulsion maneuvers were preceded by a plus-X reaction control translation to effect propellant settling, and all firings were conducted under automatic control.

Engine transient performance during all starts and shutdowns was satisfactory. During the initial firing, minor oscillations in the measured chamber pressure were observed beginning approximately 1.8 seconds after ignition. The magnitude of the oscillations was less than 30 psi peak-to-peak, and by approximately 2.1 seconds after ignition, the chamber pressure data were indicating normal steady-state operation. Similar oscillations observed during the first firing for Apollo 11 were attributed to a small amount of helium which was probably trapped in the heat exchanger after completion of bleed procedures during propellant loading.

The propellant utilization and gaging system operated satisfactorily throughout the mission. During Apollo 9, 10, and 11, the engine mixture ratio was less than expected, based on engine ground test data. Although the cause of the observed negative mixture ratio shifts have not been completely determined, the predicted flight mixture ratio for this mission was biased, based on previous flight experience, to account more closely for the expected flight mixture ratio. This biased prediction involved conducting the entire mission with the propellant utilization valve in the increase position to achieve a final propellant unbalance close to zero. Soon after ignition for the first firing, the crew moved this valve to the increase position, where it remained throughout the entire flight. The final propellant unbalance was approximately 50 pounds of oxidizer greater than the optimum quantity distribution.

## 7.9 ENVIRONMENTAL CONTROL SYSTEM

The environmental control sytem performed satisfactorily and provided a comfortable environment for the crew and adequate thermal control of the spacecraft equipment. The only anomalies noted were associated with instrumentation (see section 7.5) and clogging of both urine filters.

### 7.9.1 Oxygen Distribution

The oxygen distribution system operated normally and maintained cabin pressure at 5.0 to 5.1 psia. The overall environmental control oxygen usage rate was approximately 0.45 lb/hr, which is higher than on previous missions but is still within acceptable limits. This higher consumption is attributed to the increased purging requirements of the redesigned urine receptacle assembly and to excessive cabin leakage, which required a waiver prior to launch. However, the total indicated cryogenic oxygen usage was greater than the sum of the calculated fuel cell and environmental control usage by about 27 pounds. This discrepancy is discussed in section 14.1.7.

### 7.9.2 Thermal Control

The primary water/glycol coolant system provided adequate temperature control throughout the mission. Nearly all heat rejection was accomplished by the space radiators, with the primary evaporator activated only during launch, earth orbit, and entry. The secondary coolant system was operated only during redundant component checks and for approximately 80 minutes of evaporation before and during entry.

At about 190 hours during transearth coast, the cabin temperature decreased below the crew comfort level. The crew, following ground instructions, switched the glycol temperature control valve from automatic to manual operation and positioned the valve to increase the evaporator outlet temperature to approximately 55° F. A similar temperature increase was reflected at the suit heat exchanger and water separator, resulting in gas leaving the unit saturated to a higher water vapor level. This increased moisture content probably accounts for most of the associated condensation noted by the crew on hatches, windows, and panels.

During a special test of the high-gain antenna, the service propulsion engine was pointed toward the sun, the attitude for maximum radiator heat rejection. During this test at 193:48:00, the primary radiator heater turned on at an indicated radiator outlet temperature of minus 7° F, approximately 7° F higher than expected. This increase may have resulted from a shift in the operating band of the heater electronic control or from a difference in the glycol temperatures sensed by the heater control sensor, in the service module, and by the sensor in the command module. Inadequate flow turbulence immediately downstream of the combined radiator outlets with unequal temperatures could result in this situation. A minor control-circuit shift has no effect on system performance, while a complete failure would require switching to a redundant heater operation with separate sensors and controls. Because of difficulties in providing the necessary low radiator temperatures, preflight checkout tests do not demonstrate performance on an end-to-end basis. Consequently, some differences can be expected between flight data and temperatures determined from preflight bench checks of the controllers.

### 7.9.3 Water Management

An inline hydrogen separator was installed in the water system for the first time and successfully removed the hydrogen from the water. Some gas bubbles, probably oxygen, were noted in the hot water but were not considered objectionable. Improved gas separator cartridges also were installed on both the water gun and the food preparation unit during portions of the flight. After the cartridges were removed, little difference was noted in water quality.

After each actuation of the hot-water dispenser on the food preparation unit, the metered water flow did not shut off completely. This problem is discussed in section 14.1.15.

### 7.9.4 Waste Management

The waste management system included a redesigned urine receptacle assembly, which the crew reported was convenient to use, although care was required to prevent urine splashback. In order to avoid perturbations to passive thermal control attitudes during rest periods, the Gemini-type urine collection devices were used to store urine during these periods, rather than using the dump system. During transearth coast, the prime and backup urine filters clogged, and the urine overboard dump system was operated without a filter for the final day. This anomaly is described in section 14.1.10.

## 7.10 CREW STATION

### 7.10.1 Displays and Controls

The displays and controls in general satisfactorily supported the flight, except for the following discrepancies. The tuning fork display for the panel 2 mission clock was visibly intermittent during the prelaunch and launch phases and continuously throughout the remainder of the flight. The tuning fork display indicates that the mission clock has switched from the timing signal in the central timing equipment to an internal timing source. Section 14.1.18 contains further discussion of this malfunction. The glass faceplate of the same clock contained two cracks. This condition has occurred on clocks in several other spacecraft and is caused by stresses induced in the glass when it is bonded to the metal faceplate. New mission clocks, mechanically and electrically interchangeable with present clocks, are being developed for Apollo 14 and subsequent spacecraft.

### 7.10.2 Crew Provisions

The crew recommended that the present two-piece inflight coverall garments be retained, instead of being replaced with a one-piece item as planned. The primary advantage of the two-piece item is the capability of wearing either the jacket or trouser, or both, as required for individual comfort. In addition, the crew recommended an additional set of inflight coverall garments be stowed for personal comfort and hygiene, since the original set can become very dirty late in the mission.

The metal window shades were difficult to fit and secure, with windows 1 and 5 reported to be the most difficult. The shades for windows 1, 2, 4, and 5 are installed into the window frame by slipping one end under two finger clips and rotating the swivel latches over the shade rim to secure it in place. To allow proper engagement in flight, the crew pried the finger clips with the adjustable wrench to increase the clearance for shade insertion and adjusted the length to the swivel latches. During ground and altitude chamber test checks, the crew had properly fit the window shades with little effort. A modification, now being implemented for Apollo 13, deletes the finger clips and provides spring-loaded latches in a three-point engagement.

## 7.11 CONSUMABLES

The command and service module consumables usage during the Apollo 12 mission were well within the red line limits and, in all cases except one, differed no more than 5 percent from the predicted limits.

### 7.11.1 Service Propulsion Propellant

Service propulsion propellant usage was within 1 percent of the preflight estimate for the mission. The propellant unbalance was less than 50 pounds after the final firing and is the lowest unbalance experienced during any Apollo mission. In the following table, the loadings were calculated from gaging system readings and measured densities at lift-off.

| Conditions | Actual usage, lb Fuel | Oxidizer | Total | Preflight planned usage, lb |
|---|---|---|---|---|
| Loaded | 15,728 | 25,089 | 40,817 | 40,817 |
| Consumed | | | 37,080 | 36,675 |
| Remaining at command module/service module separation | | | 3,737 | 4,142 |

### 7.11.2 Reaction Control Propellant

Service module.- Consumption of service module reaction control propellant was about 28 percent greater than predicted. The increased usage resulted partly from operating at a 0.5-deg/sec maneuver rate with the digital autopilot early in the mission, instead of the usual 0.2 deg/sec rate. The remainder of the greater than predicted consumption was used for unplanned landmark tracking activities during lunar orbit. Despite this increased consumption, the quantity of propellant remaining always remained well above the red line limit. The usages listed in the following table were calculated from telemetered helium-tank-pressure data and were based on the relationship of the pressure, volume, and temperature.

| Condition | Propellant, lb | | | Preflight Planned propellant, lb |
|---|---|---|---|---|
| | Fuel | Oxidizer | Total | |
| Loaded | | | | |
| Quad A | 111 | 225 | | |
| Quad B | 110 | 225 | | |
| Quad C | 110 | 224 | | |
| Quad D | 110 | 225 | | |
| Total | 441 | 899 | 1341 | 1340 |
| Consumed | 318 | 637 | 955 | 680 |
| Remaining at command module/service module separation | 123 | 263 | 386 | 660 |

Command module.- The actual usage of command module reaction control propellant agreed with predicted usage to within 17 percent. The calculated quantities listed in the following table are based on pressure, volume, and temperature relationships, and an average mixture ratio of 1.85.

| Condition | Actual quantities, lb | | | Preflight planned quantities, lb |
|---|---|---|---|---|
| | Fuel | Oxidizer | Total | |
| Loaded (usable) | | | | |
| System 1 | 40.6 | 63.6 | | |
| System 2 | 40.6 | 63.6 | | |
| Total | 81.2 | 127.2 | 208.4 | 208.6 |
| Consumed | | | | |
| System 1 | 12 | 23 | 35 | 40 |
| System 2 | 0 | 0 | 0 | 0 |
| Remaining at main parachute deployment | | | | |
| System 1 | 28.6 | 40.6 | 69.2 | |
| System 2 | 40.6 | 63.6 | 104.2 | |
| Total | 69.2 | 104.2 | 173.4 | |

### 7.11.3 Cryogenics

The oxygen and hydrogen usages were within 8 percent of those predicted. Usages listed in the following table are based on quantity data transmitted by telemetry.

| Condition | Hydrogen, lb Actual | Planned | Oxygen, lb Actual | Planned |
|---|---|---|---|---|
| Available at lift-off | | | | |
| Tank 1 | 26.5 | | 319.0 | |
| Tank 2 | 27.3 | | 316.0 | |
| Total | 53.8 | 53.2 | 635.0 | 600.0 |
| Consumed | | | | |
| Tank 1 | 21.7 | | 248.0 | |
| Tank 2 | 22.5 | | 237.0 | |
| Total | 44.2 | 45.0 | 485.0 | 445.0 |
| Remaining at command module/ service module separation | | | | |
| Tank 1 | 4.8 | | 71.0 | |
| Tank 2 | 4.8 | | 79.0 | |
| Total | 9.6 | 8.2 | 150.0 | 155.0 |

### 7.11.4 Water

Predictions concerning water consumption in the command and service modules are not made because the water system has an initial charge of potable water at lift-off and more than ample water for environmental control and crew consumption is generated by fuel-cell reaction. The water quantities loaded, consumed, produced, and expelled during the mission are shown in the following table.

| Condition | Quantity, lb |
|---|---|
| Loaded | |
| Potable water tank | 20.6 |
| Waste water tank | 27.9 |
| Produced inflight | |
| Fuel cells | 390.2 |
| Lithium hydroxide, metabolic | 45.5 |
| Dumped overboard (including urine)[a] | 398 |
| Evaporated up to command module/ service module separation | 8.6 |
| Remaining at command module/service module separation | |
| Potable water tank | 36.4 |
| Waste water tank | 41.9 |

[a] This parameter can only be estimated from flight data.

## 8.0 LUNAR MODULE PERFORMANCE

Performance of the lunar module systems is discussed in this section. The thermal control system performed as intended and is not discussed further, and this section included a discussion of the performance of the extravehicular mobility unit. Discrepancies and anomalies in lunar module systems are generally mentioned in this section but are discussed in greater detail in the anomaly summary, sections 14.2 and 14.3, the late latter comprising government furnished equipment.

## 8.1 STRUCTURAL AND MECHANICAL SYSTEMS

The structural analysis was based on guidance and control data, cabin pressure measurements, command module acceleration data, photographs, and crew comments.

Based on measured command module accelerations and on simulations using actual launch wind data, lunar module loads were within structural limits during earth launch and translunar injection. Loads during both dockings and the three docked service propulsion maneuvers were also within structural limits.

The sequence films from the onboard camera showed no evidence of structural oscillations during lunar touchdown, and crew comments agree with this assessment. Flight data from the guidance and propulsion systems were used in performing engineering simulations of the touchdown phase (section 4.2). As in Apollo 11, the simulations and photographs indicate that landing gear stroking was minimal and that external loads were well within design values.

During his initial egress, the Commander's life support package tore a portion of the thermal shielding on the forward hatch. While this tear did not compromise the thermal integrity of the spacecraft, the possibility of contact on future missions could represent a hazard to suit pressure integrity. This anomaly is discussed further in section 14.2.6.

The deployment ring for the external equipment storage compartment failed to operate properly, and the Commander was required to deploy the compartment door by pulling on the lanyard attached to the ring. This discrepancy is discussed in section 14.2.5.

## 8.2 ELECTRICAL POWER

Electrical power system performance was satisfactory throughout the mission. The descent batteries supplied 1023 ampere-hours of power from a nominal total capacity of 1600 ampere-hours, and at final docking, the ascent batteries had delivered 230 ampere-hours from a nominal total capacity of 592 ampere-hours. All power switchovers were accomplished as required, and parallel operation of the descent and ascent batteries was within acceptable limits. The bus voltage during powered-up operations was maintained above 28.6 V dc. The maximum electrical load, 77 amperes, was momentarily observed during the powered descent maneuver. The total battery energy usage throughout lunar module flight followed preflight predictions to within 1 percent.

## 8.3 COMMUNICATIONS EQUIPMENT

Performance of the communications systems was satisfactory. However, the crew reported that VHF voice communications between the two spacecraft were unacceptable during the ascent, rendezvous, and docking portions of the mission. Section 14.1.19 includes a detailed discussion of this problem.

During the first extravehicular period, the S-band erectable antenna was operationally deployed for the first time in the Apollo program. Following ingress, the antenna was used for S-band communication until approximately 30 minutes prior to ascent. This antenna provided the predicted gain increase and enabled use of the low power S-band mode during the lunar sleep period.

During the entire extravehicular activity, the lunar module relay mode provided good voice and telemetry data transmission. However, a tone, accompanied by random impulse noise, was present intermittently for approximately 2 hours during the first extravehicular excursion. The tone, but without the noise, was present for approximately 12 seconds during the second extravehicular operation. Postflight tests revealed the left microphone amplifier in the Commander's communications carrier had been intermittent. The amplifier failure has not been correlated to the audible tone, but a random noise, similar to that heard during extravehicular activity, was detected whenever the microphone was intermittent. Because the communications carrier has redundant microphones and amplifiers, no loss of communications was associated with the amplifier failure. See section 14.1.19 for further discussion of this problem. As experienced on Apollo 11, an intermittent uplink voice echo was noted during extravehicular activity. The echo was of a lower level than experienced on Apollo 11, and communications were considered to have been satisfactory.

Reception from the color television camera was nominal until the camera vidicon tube was damaged by either a direct or reflected image of the sun after approximately 40 minutes of operation during the first extravehicular period. See section 14.3.1 for a more detailed discussion.

### 8.4 RADAR

Landing radar performance during powered descent was normal. Acquisition of range and velocity occurred at 41,438 and 40,100 feet, respectively. Two brief dropouts occurred at low altitude during the hovering phase. The first dropout appeared at approximately 234 feet slant range and the second at 44 feet slant range. Analysis revealed the spacecraft was undergoing a translation to the right at these times, and dropouts are expected under these conditions because of a zero Doppler effect in either beam 1 or 2. Three abnormally high data points appeared just prior to touchdown. At altitudes below 50 feet, the range and velocity trackers are operating on highly attenuated signals resulting from the high discrimination of the receiver audio amplifiers to the low frequency signals at these trajectory conditions. Since the trackers are approaching signal dropout, the velocity trackers are particularly vulnerable to locking up on moving dust and debris generated by exhaust plume impingement on the lunar surface. Also, under these conditions, the range tracker is vulnerable to locking up at higher frequencies because of terrain features appearing in the range-beam side lobes.

Rendezvous radar performance was normal in all respects. Just prior to docking, a loss of a radar "data good" indication occurred at a range of 150 feet, and was earlier than expected. No further rendezvous radar data were required, so the crew opened the associated circuit breakers. No anomalies are indicated from the data, and the loss of the "data good" indication was caused by a brief drop in signal strength as a result of rapid attitude changes.

### 8.5 INSTRUMENTATION

Performance of the instrumentation system was satisfactory. The only unexplained master alarm occurred just prior to ascent engine ignition. Any of the non-latching caution and warning inputs could have been subjected to a momentary out-of-tolerance condition sufficient to cause a master alarm without being detected by the crew or the ground. Sections 14.2.3 and 14.2.7 contain discussions of a carbon-dioxide sensor malfunction and an early indication from the fuel-quantity low-level sensor respectively.

### 8.6 GUIDANCE AND CONTROL

Guidance and control system performance was satisfactory throughout the mission. This section describes overall system operation and highlights the ascent and rendezvous portions of flight. A discussion of guidance and control system performance during powered descent and landing is contained in section 4.2.

Because of the lightning encountered during launch, the primary guidance computer was powered up and verified ahead of schedule early in translunar coast. An erasable memory dump was performed which indicated that no adverse effects had been experienced. The power-up sequence in lunar orbit prior to unlocking was normal and proceeded with no difficulty. The inertial measurement unit was aligned as in previous missions by transferring command module platform gimbal angles across the structural interface between the two spacecraft and by taking into account the relative orientation of the two vehicles and the roll-axis misalignment observed on the docking ring scale. For the first time in Apollo, a drift check was then performed utilizing a new technique which compared the rotation vectors measured by each platform during successive attitude maneuvers and used the vector differences to calculate any misalignment. A gyro drift measurement was also obtained from an optical alignment performed after undocking. Table 8.6-I contains the results of inflight and lunar surface alignments performed during the mission. Table 8.6-II contains a guidance systems alignment comparison.

The crew reported observing small attitude display changes at times when switching the flight-director-attitude-indicator drive source between primary and abort guidance system attitude references. The changes occurred both immediately and at later times following alignments. The observed changes are a normal characteristic for this type of mechanization and result from a combination of errors from the following sources.

| Source | Specification error, deg Roll | Yaw | Pitch |
|---|---|---|---|
| Platform/gimbal angle sequence transformation assembly interface | ±0.3 | ±0.3 | ±0.3 |
| Gimbal angle sequence transformation assembly static accuracy | ± 0.75 | ±1.1 | ±1.75 |
| Abort guidance system signal accuracy | ±0.5 | ±0 .5 | ±0.5 |

## TABLE 8.6-I.- INFLIGHT AND LUNAR SURFACE ALIGNMENT DATA

| Time, hr:min | Type alignment | Alignment mode Option[a] | Alignment mode Technique[b] | Telescope detent[c] /star used | Star angle difference, deg | Gyro torquing angle, deg X | Gyro torquing angle, deg Y | Gyro torquing angle, deg Z | Gyro drift, mERU X | Gyro drift, mERU Y | Gyro drift, mERU Z |
|---|---|---|---|---|---|---|---|---|---|---|---|
| 104:52 | | Docked | alignment | | | -0.250 | -0.360 | +0.050 | —[d] | —[d] | —[d] |
| 108:11 | | Docked | alignment | | | -0.045 | -0.035 | -0.092 | 0 .9 | 0.7 | 1.8 |
| 108:48 | P52 | 3 | NA | 2/13; 2/12 | 0.02 | +0.018 | -0.002 | -0.069 | 0.3 | 0.0 | 1.2 |
| 110:46 | P57 | 3 | 1 | NA | 0.07 | -0.011 | +0.064 | -0.054 | 0.4 | 2.2 | 1.8 |
| 110:54 | P57 | 3 | 2 | 1/15; 2/00 | 0.01 | +0.027 | -0.017 | -0.045 | — | — | - |
| 111:22 | P57 | 3 | 2 | 1/16; 6/17 | 0.02 | +0.034 | +0.036 | +0.019 | — | — | - |
| 139:26 | P57 | 4 | 3 | 1/16; -/- | 0.04 | +0.001 | +0.057 | +0.033 | — | — | - |
| 141:29 | P57 | 4 | 3 | 1/16; -/- | 0.04 | -0.023 | +0.004 | +0.015 | 0.7 | 0.1 | 0.5 |
| 142:23 | P52 | 3 | NA | 2/12; 2/13 | 0.01 | +0.008 | +0.010 | -0.046 | | | |

[a] 1 - Preferred; 2 - Nominal; 3 - REFSMMAT; 4 - Landing site.
[b] 0 - Anytime; 1 - REFSMMAT plus g; 2 - Two bodies; 3 - One body plus g.
[c] 1 - Left front; 2 - Front; 3 - Right front; 4 - Right rear; 5 - Rear; 6 - Left rear.
[d] Not torqued.

Star names:
13 Capella
12 Rigel
15 Sirius
00 Pollux
16 Procyon
17 Regor

## TABLE 8.6-II.- GUIDANCE SYSTEMS ALIGNMENT COMPARISON

Primary minus abort system

| Time of alignment | Alignment error (degrees) X | Y | Z |
|---|---|---|---|
| | Before powered descent | | |
| 106:11:48 | -0.011 | 0.013 | -0.008 |
| 106:48 :26 | * | * | * |
| 108:38:57 | -0.020 | 0.001 | -0.009 |
| 108:39:09 | -0.025 | 0 | 0.017 |
| 110:16:54 | -0.005 | -0.014 | -0.010 |
| | Lunar surface | | |
| 111:33:34 | 0.004 | -0.024 | 0.001 |
| 139:36:11 | -0.013 | 0 | 0 |
| 139:50:27 | -0.013 | 0.035 | -0.001 |
| 141:31:53 | -0.002 | -0.005 | 0.004 |
| | After docking | | |
| 147:22:48 | -0.047 | 0.005 | 0.009 |

*Data not available.

The digital autopilot was used almost exclusively for attitude control during the mission, and performance was normal throughout. Spacecraft response during descent, ascent, and reaction control system maneuvers was as expected. Although the crew reported an unexpected amount of reaction control system activity during descent, data indicate normal duty cycles (see section 4.2). The crew concern appears to have resulted from a software discrepancy in preflight lunar module simulations.

System operation after lunar touchdown was nominal and proceeded according to schedule. The landing coordinates, as obtained from lunar surface alignments and rendezvous radar data, are discussed in section 4.3 and are shown in figure 4-11.

The ascent trajectory was very close to nominal. A procedural error involving late actuation of the engine-arm switch resulted in a 32.5-ft/ sec overburn, which was immediately trimmed with the reaction control system. The effect of accelerometer bias errors in the primary guidance system is indicated in table 8.6-III, which is a comparison of insertion conditions as measured onboard and by the ground.

**TABLE 8.6-III.- LUNAR ORBIT INSERTION CONDITIONS**

| Altitude, Source | Vertical velocity, feet | Horizontal velocity, ft/sec | ft/sec |
|---|---|---|---|
| Primary guidance | 62,677 | 41.6 | [a]5530 |
| Abort guidance | 61,504 | 38.6 | 5536 |
| Network tracking | 62,380 | 41.4 | 5537 |

[a] Four ft/sec of the difference between primary and abort guidance systems is due to a bias error in the primary guidance Z pulse integrating pendulous accelerometer.

The ascent and rendezvous profiles were very similar to those for Apollo 11, with the exception that the abort guidance system was planned to be used independently of the primary system. This change was accommodated by independently maintaining the abort guidance system state vector during rendezvous while manual inputting of radar data. The ascent preparation sequence was nominal and closely followed the flight plan. Figure 8.6-1 is a time history of attitude rates at lift-off. Because no data dropouts occurred, as in Apollo 11, an attitude-rate analysis of this phase was possible for the first time. The transients were well within the controllability limit and indicated reasonable agreement with preflight simulations.

Primary guidance solutions were used exclusively during rendezvous. See table 5-VII for a comparison of the various available solutions. The crew reported an excessive workload was involved in maintaining the abort guidance system independent of the primary system throughout rendezvous. The only discrepancy reported during the rendezvous was procedural and occurred when a radar update in range and range rate was loaded in an incorrect sequence. The out-of-sequence updating severely degraded the abort guidance system state vector and caused the maneuver solution to be incorrect. Thereafter, the abort guidance system was externally targeted using the primary guidance maneuver solution for maneuver backup purposes.

Inertial measurement unit operation was satisfactory throughout the mission. Accelerometer bias had been extremely stable in the period from power-up through landing; however, all accelerometers exhibited a step change across the power-down and power-up sequences on the lunar surface, as shown in table 8.6-IV. Although the measurements of total bias made on the surface contain errors as a result of the uncertainties in magnitude and direction of gravity, shifts in the measured values are detectable. The step changes were minor and within system operating limits.

The guidance computer performed as expected throughout the descent and ascent phases. No alarms were experienced during powered descent, indicating that software improvements made as a result of the Apollo 11 master alarms were successful.

Alignment optical telescope performance was excellent. Because of the more westerly location of the landing site and the sun and earth positions with respect to the telescope lines of sight, more of the detents were usable than on the previous mission.

The abort guidance system was used solely in a backup role throughout the mission. The results of the inflight and lunar surface calibrations and other inertial component performance measurements are shown in table 8.6-V and 8.6-VI and indicate excellent performance throughout.

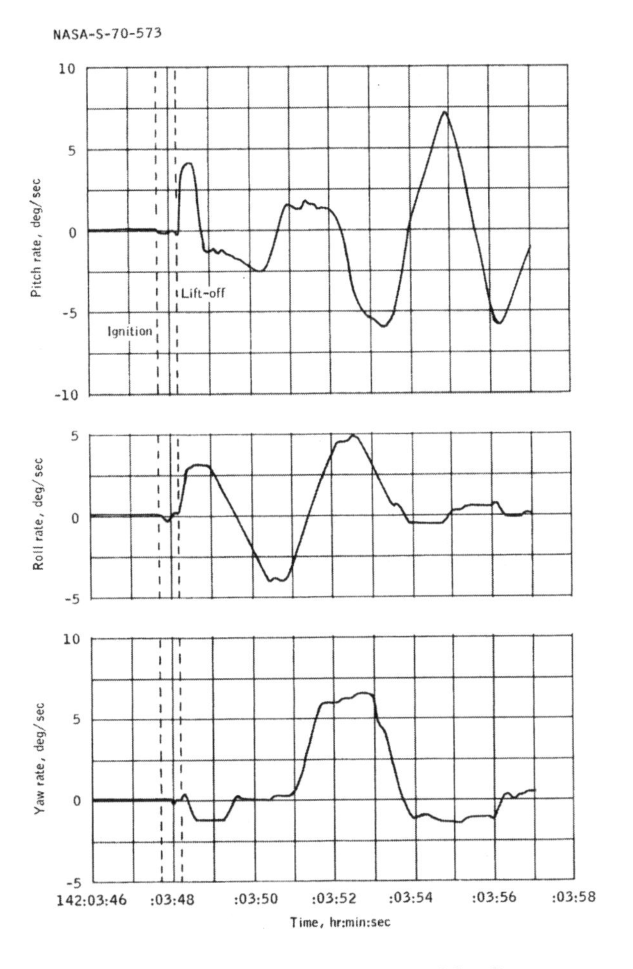


Figure 8.6-1 Attitude rates at lunar lift-off

## TABLE 8.6-IV.- INERTIAL COMPONENT PREFLIGHT HISTORY - LUNAR MODULE

(a) Accelerometers

| Error | Sample mean | Standard deviation | Number of samples | Countdown value | Flight load | Inflight performance | | | | | |
|---|---|---|---|---|---|---|---|---|---|---|---|
| | | | | | | Power-up to 106:43 | Update (106:43) | Landing to power-down | Surface power-up to lift-off | Update (143.45) | 143:45 to rendezvous |
| X - Scale factor error, | | | | | | | | | | | |
| ppm ......... | -649 | 18 | 4 | _640 | _660 | — | — | — | — | — | — |
| Bias, cm/sec² ..... | -0.39 | 0.02 | 4 | -0.37 | -0.38 | -0.33 | -0.33 | -0.40 | -0.10 | -0.15 | -0,17 |
| Y - Scale factor error, | | | | | | | | | | | |
| ppm ......... | -681 | 72 | 4 | -727 | -720 | — | — | — | — | — | — |
| Bias, cm/sec² ..... | 0.03 | 0.01 | 4 | 0.03 | 0.02 | 0.01 | — | 0.05 | 0.20 | 0.20 | 0.18 |
| Z - Scale factor error, | | | | | | | | | | | |
| ppm ......... | -885 | 42 | 4 | -943 | -890 | — | — | — | — | — | — |
| Bias, cm/sec² ..... | 0.60 | 0.05 | 4 | 0.63 | 0.62 | 0.68 | 0.68 | 0.73 | 0.34 | 0.39 | 0.42 |

(b) Gyroscopes

| Error | Sample mean | Standard deviation | Number of samples | Countdown value | Flight load | inflight performance |
|---|---|---|---|---|---|---|
| X - Null bias drift, mERU ...... | -1.0 | 0.3 | 5 | -1.3 | 0.1 | 0.6 |
| Acceleration drift, spin reference axis, mERU/g ......... | -1.3 | 1.4 | 4 | -0.4 | -2.0 | — |
| Acceleration drift, input axis, mERU/g ......... | 10.6 | 6.5 | 4 | 14.0 | 7.0 | — |
| Y - Null bias drift, mERU ...... | 0.7 | 1.0 | 5 | -0.2 | 0.8 | 0.8 |
| Acceleration drift, spin reference axis, mERU/g ......... | 4,1 | 1.4 | 4 | 5.3 | +4.0 | — |
| Acceleration drift, input axis, mERU/g ......... | -16.0 | 6.8 | 4 | -23.3 | -15.0 | — |
| Z - Null bias drift, mERU ...... | 2.8 | 0.9 | 5 | 3.3 | 3.0 | 1.3 |
| Acceleration drift, spin reference axis, mERU/g ......... | -0.3 | 4.2 | 4 | -2.6 | -2.0 | — |
| Acceleration drift, input axis, mERU/g ......... | 10.8 | 4.8 | 4 | 12.8 | 13.0 | — |

## TABLE 8.6-V.- ABORT GUIDANCE SYSTEM PREINSTALLATION CALIBRATION DATA

| Accelerometer bias | Sample mean, mg | Standard deviation, mg | Number of samples | Final calibration value, mg | Flight load value, mg |
|---|---|---|---|---|---|
| X | 474 | 25.6 | 12 | 462 | 451 |
| Y | 138 | 27.5 | 12 | 119 | 119 |
| Z | -83 | 10.7 | 12 | -79 | -71 |

| Accelerometer scale factor | | Standard deviation, ppm | Number of samples | Final calibration value, ppm | Flight load value, ppm |
|---|---|---|---|---|---|
| X | 35 | 7 | 13.4 | 1282 | |
| Y | 29 | 7 | -1589 | -1637 | |
| Z | 32 | 7 | -2265 | -2314 | |

| Gyro scale factor | Sample mean, ppm | Standard deviation, ppm | Number of samples | Final calibration value, ppm | Flight load value, ppm |
|---|---|---|---|---|---|
| X | 610 | 10.8 | 12 | 615 | 615 |
| Y | 3282 | 8.1 | 12 | 3294 | 3294 |
| Z | 2930 | 10.0 | 12 | 2941 | 2941 |

| Gyro fixed drift deg/hr | Sample mean, deg/hr | Standard deviation, samples | Number of deg/hr | Final calbration value, deg/hr | Flight load value, |
|---|---|---|---|---|---|
| X | 0.014 | 0.062 | 12 | 0.06 | 0.06 |
| Y | -0.096 | 0.054 | 12 | -0.16 | -0.16 |
| Z | -0.002 | 0.048 | 12 | -0.07 | -0.07 |

| Gyro spin axis mass unbalance | Sample mean deg/hr | Standard deviation, deg/hr | Number of samples | Final calibration value, deg/hr | Flight load value, deg/hr |
|---|---|---|---|---|---|
| X | 0.154 | 0.117 | 12 | 0.03 | 0.03 |

**TABLE 8.6-VI.- ABORT GUIDANCE SYSTEM GYRO CALIBRATION DATA**

| | X, deg/hr | Y, deg/hr | Z, deg/hr |
|---|---|---|---|
| Preinstallation calibration | +0.06 | -0.16 | -0.07 |
| Final earth prelaunch calibration | -0.27 | -0.31 | -0.06 |
| Inflight calibration | -0.04 | -0.19 | 0 |
| First lunar surface calibration | -0.19 | -0.28 | +0.11 |
| Third lunar surface calibration | -0.20 | -0.31 | +0.05 |

## 8.7 REACTION CONTROL

Reaction control system performance was normal in all respects. Onboard measurement of propellant consumption through ascent stage jettison was 315 pounds, compared with the predicted value of 305 pounds. Reaction control system interconnect operation was satisfactory during the ascent maneuver; however, the indicator for the system A main shutoff valve remained in the valve-closed position after the valves had been initially commanded open. This indicator operated normally when the valves were recycled (section 8.11.1 has a more complete discussion).

The thrust-chamber pressure switch on the quad 4 side-firing engine failed in the closed position for about 2 minutes during powered descent. This switch, which also failed closed several times during ascent, was slow in opening on all firings after undocking. However, engine performance was nominal at these times. This type of failure, noted on all previous manned lunar modules, is attributed to particulate contamination of the switch. The only consequence of such a of failure is that a failed-off engine cannot be detected from instrumentation sources.

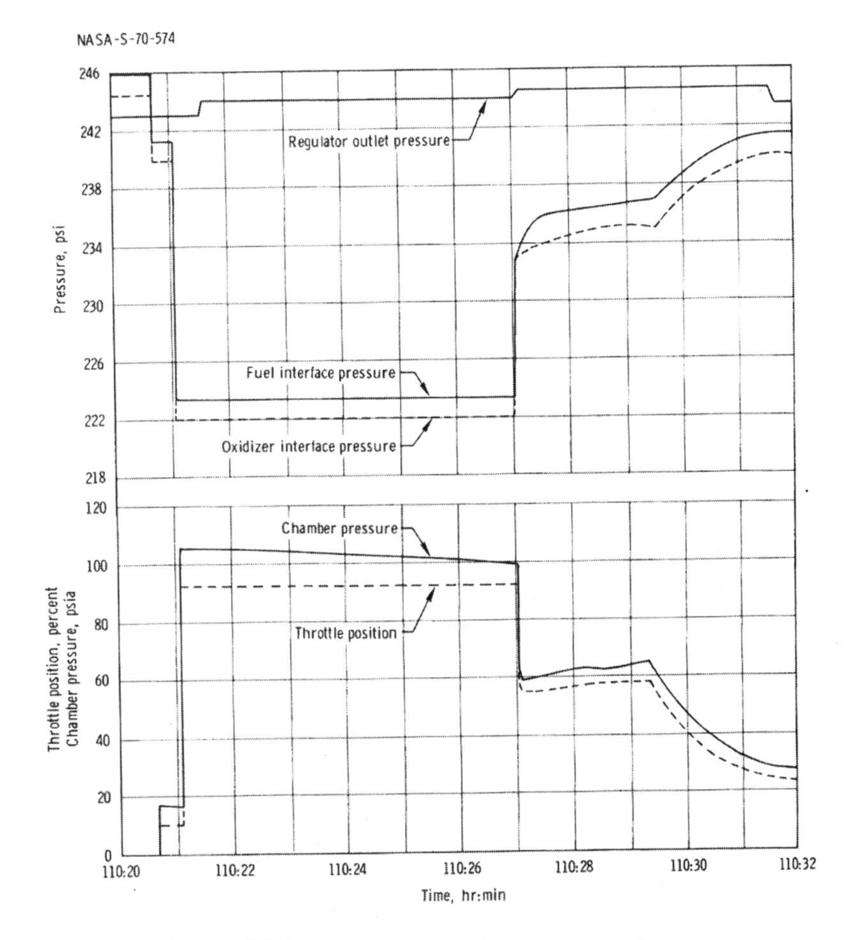


Figure 8.8-1 Descent propulsion system performance

## 8.8 DESCENT PROPULSION

Descent propulsion system operation, including engine starts and throttle response, was normal.

### 8.8.1 Inflight Performance

The descent propulsion system performed normally during the 29-second descent orbit insertion maneuver. The powered descent firing lasted 717 seconds, and the system pressures and throttle settings are presented in figure 8.8-1. The data curve has been smoothed and does not reflect the numerous throttle changes made during the final descent. During powered descent, the oxidizer interface pressure appeared to be oscillating as much as 59 Psi peak to peak. These oscillations were evident throughout the firing but were most prominent at about 55- to 60-percent throttle.

Oscillations of this type were also observed during the Apollo 11 descent. After the Apollo 11 flight, it was determined that the oscillations resulted from the instrumentation configuration and were not inherent in the system. Engine performance and operation were not affected in either flight.

### 8.8.2 System Pressurization

The oxidizer tank ullage pressure decayed from 94 to 60 psia during the period from lift-off to second activation of the system at about 90 hours. During that period, the fuel tank ullage pressure decreased from 128 to 105 psia. These decays were within the expected range for helium absorption into the propellants.

The measured pressure profile of the supercritical helium tank was within acceptable limits. The pressure rise rates on the ground and in flight were 8.0 and 6.1 psi/hr, respectively.

The procedure for venting the propellant tanks .after landing was changed from Apollo 11, during which a freeze-up of the line to the supercritical helium tank occurred (reference 9). The supercritical helium tank was isolated prior to the venting, which was then accomplished successfully, and the helium tank was subsequently vented 21 minutes before ascent stage lift-off. During the lunar stay period, the pressure rise rate was 4.9 psi/hr.

### 8.8.3 Gaging System Performance

The descent propellant gages indicated expected quantities throughout lunar module flight. The two fuel probe measurements agreed to within approximately 1 percent throughout powered descent, and the difference remained relatively constant. The oxidizer probe measurements diverged with time until mid-way through the firing, although the difference was only 1 percent. After that point, the difference remained constant. The slight divergence was probably caused by oxidizer flowing from tank 2 to tank 1 through the propellant balance line, as a result of an offset in the vehicle center of gravity.

The low-level light came on at 110:31:59.6 (after 681.5 seconds of firing time) and was apparently triggered by the fuel tank 2 point sensor, which had the lowest reading. This light indicated that 5.6 percent fuel quantity remained. This quantity is equivalent to approximately 113 seconds of total firing time remaining to propellant depletion, based on the sensor location. Postflight data for the gaging system probe, however, indicate that the propellant readings were oscillating from 1.5 to 2.0 percent peak-to-peak about the mean reading. This oscillation was indicative of propellant slosh, which could cause a premature low-level indication. Based on the mean propellant reading of 6.7 percent quantity remaining, the sensor should have been activated approximately 25 seconds later than indicated. Engine shutdown occurred 35.5 seconds after the low-level signal, and the associated firing time remaining should have been 77.5 seconds. However, the low-level indication was received early and a firing time of 103 seconds to fuel tank 2 depletion actually remained. Even with the apparent slosh-induced error, the difference between the continuous probe reading and the low-level light indication was within the expected accuracy of the gaging system.

## 8.9 ASCENT PROPULSION

The ascent propulsion system performed satisfactorily during the 425-second ascent maneuver (engine on to engine off). Helium regulator outlet pressure dropped from a level of 189 psia to the expected value of approximately 185 psia at engine ignition. However, both measurements for helium regulator outlet pressure showed oscillations throughout the firing with respective maximum recorded amplitudes of 6 and 19 psi peak to peak. Similar oscillations, with approximately the same amplitudes, were observed from Apollo 10 data, as well as oscillations with smaller amplitudes during ground testing. It was concluded from the evaluation of Apollo 10 data that a portion of the oscillation magnitude was attributable to certain characteristics of the pressure transducers. No degradation in system performance from these pressure oscillations has been noted for either Apollo 10 or 12.

Table 8.9-I is a summary of actual and predicted performance parameters during the ascent-engine firing, which was approximately 6 seconds shorter than expected, based on preflight performance estimates. The shorter firing time may be attributed to a combination of lower-than-expected vehicle weight, higher-than-predicted engine performance, and a greater-than-expected impulse from "fire-in-the-hole" effects. A more detailed reconstruction of data will be presented in a supplemental report (see appendix E).

During the coast period following ascent, the oxidizer system pressure dropped in a manner and magnitude similar to that observed on Apollo 11. This phenomenon is discussed in reference 9 and had no apparent effect on spacecraft performance or crew safety.

**TABLE 8.9-I.- STEADY-STATE PERFORMANCE**

| Parameter | 10 seconds after ignition | | 400 seconds after ignition | |
|---|---|---|---|---|
| | Predicted [a] | Measured [b] | Predicted [a] | Measured [b] |
| Regulator outlet pressure, Asia .... | 184 | 184[c] | 184 | 184[c] |
| Oxidizer bulk temperature, °F ..... | 69.9 | 68.5 | 69.0 | 67.8 |
| Fuel bulk temperature, °F ....... | 69.7 | 68.5 | 69.5 | 68.5 |
| Oxidizer interface pressure, psia ... | 171.1 | 168.0 | 170.2 | 167.5 |
| Fuel interface pressure, psia ..... | 170.6 | 167.5 | 169.8 | 166.7 |
| Engine chamber pressure, psia ..... | 123.0 | 120.0 | 122.7 | 119.5 |
| Mixture ratio ............. | 1.611 | — | 1.602 | — |
| Thrust, lb .............. | 3495 | — | 3460 | — |
| Specific impulse, sec ......... | 309.5 | — | 309.2 | — |

[a] Preflight prediction based on acceptance test data and assuming nominal system performance.
[b] Actual flight data with known biases removed.
[c] These values are approximate due to oscillations noted in text.

## 8.10 ENVIRONMENTAL CONTROL SYSTEM

The environmental control system satisfactorily supported all lunar module operations throughout the mission. Although water in the suit loop and an erratic carbon dioxide sensor have been identified as anomalies, overall performance was nominal and lunar module operations were not compromised.

On the lunar surface, the cabin was depressurized through the forward dump valve without a cabin-gas bacteria filter installed as modified for this mission. Cabin pressure decreased rapidly, as predicted, and the crew was able to open the hatch 3 minutes after actuation.

Prior to the first extravehicular activity, the crew reported free water in the suit inlet umbilicals. After the mission, the umbilical assemblies were tested under flight conditions, and no condensation was observed. During postflight tests, condensate was observed to bypass the water separators because the separator rotational velocity was excessive as a result of the suit-circuit flow being higher than the specification value. For Apollo 13 and thereafter, an orifice will be placed in the suit circuit to reduce the flow and should decrease the separator velocity to within expected ranges. Further details are given in section 14.2.2

The Apollo 11 crew had reported that sleep was difficult because of a cold environment. This condition was remedied for Apollo 12 through the use of hammocks and through procedural changes which eliminated pre-chilling of the crew prior to the beginning of their sleep period. Although the crew reported they were comfortable during the sleep period on the lunar surface, they were awakened on occasion by an apparent change in the sound pitch produced from the water/glycol pump installation. This pump package is mounted on a bulkhead in the aft cabin floor area which is not generally subjected to significant variations in cabin temperature or pressure. All pump performance data, including temperature, line pressure, and input voltage, appear normal during the sleep period, indicating the pump frequency could not have varied perceptibly. Cabin temperature and pressure were also essentially constant during this period. The only explanation for the change in pitch, while unlikely, is that the fluid lines and supporting structure near and downstream from the pump experienced physical changes which altered the vibrational harmonics sufficient to produce, on occasion, detectable changes in pitch frequency. Because all pump parameters indicated normal operation, no system modifications are required. However, reports on past flights of an annoying noise level in the cabin has prompted a modification to the plumbing for future flights which significantly reduces noise and which will probably eliminate any pitch variations from surrounding structure.

Behind the moon during the second revolution after lunar lift-off, erratic fluctuations in the carbon dioxide partial-pressure sensor activated the caution-and-warning system, and the crew selected the secondary lithium hydroxide cartridge. The secondary cartridge also exhibited erratic indications. This condition was expected, because a similar problem was observed during Apollo 11 and was determined to be the result of free water from the water separator drain tank being introduced into the sensor casing. The sensor line will be relocated to prevent recurrence of this problem, as discussed in section 14.2.3.

## 8.11 CREW STATION

### 8.11.1 Displays and Controls

The displays and controls functioned satisfactorily in all but the following areas.

The main shutoff valve flag indicator for the system-A reaction control system did not indicate properly when the valve was commanded open; however, telemetry data showed that the valve had opened, thus indicating faulty flag operation. This indicator had exhibited sticky operation during a ground test, and the discrepancy is generic to flag indicators.

After lunar lift-off, the exterior tracking light operated normally during the first darkness pass but did not operate during the second darkness pass. The light switch was cycled, and telemetry indicated that power consumption was normal after the failure occurred. The power indication confirmed normal operation of the power supply and isolated the failure to the high-voltage section of the light. Section 14.2.4 contains further details of this problem.

The docking hatch floodlight switch failed to turn off the floodlights after the first lunar module checkout. The crew checked the switch manually, and it performed correctly. An improper adjustment between the switch and the hatch was the likely cause of the problem, and an improved installation procedure will be implemented for future missions. For further discussion of this problem, see section 14.2.1.

### 8.11.2 Crew Provisions

When the Commander attempted to zero the portable life support system feedwater bag scale, the zero adjustment nut came off. The nut was reinstalled with difficulty, and the feedwater was successfully weighed. If the scale is required for future missions, the zero-adjustment screw will be lengthened and the end peened to retain the adjustment nut.

The lunar equipment conveyor satisfactorily transferred equipment into the lunar module, although a considerable amount of lunar dust was picked up during the operation. One problem with the lunar equipment conveyor occurred at initial deployment, when the retaining pin on the strap slipped out of the conveyor stirrup. The Lunar Module Pilot corrected this condition by replacing the strap through the stirrup, and no further problems occurred. The retaining pin will be modified to preclude this problem on future missions.

## 8.12 EXTRAVEHICULAR MOBILITY UNIT

Performance of the extravehicular mobility unit was excellent during both extravehicular periods. After a brief acclimation phase, crew mobility with the extravehicular mobility unit was excellent in the 1/6-g lunar environment. Balance, stability, and movement were essentially the same as for Apollo 11. The metabolic rates and the oxygen and feedwater consumptions were lower than predicted (table 8.12-I), as also observed during Apollo 11. The crewmen remained comfortable, and only an occasional opening of the portable life support system diverter valve beyond minimum cooling was required for crew comfort.

Preparations for the first extravehicular activity proceeded rapidly, with only minor problems. On the Lunar Module Pilot's portable life support system, the tab for the lithium hydroxide canister cover lock apparently did not snap into the locked position while closing. Although the cover was locked, the Lunar Module Pilot manually verified tab locking as a precautionary measure. The failure to audibly lock into the detent position was undoubtedly caused by the locking ring and the dish having a slight misalignment, which did not actually prevent detent locking. The misalignment has been duplicated on identical hardware, with locking characteristics similar to those observed, but is not a problem. A concentricity check will be made on all future flight canisters.

Two delays during preparation for the first extravehicular activity were caused by deviating from the checklist. The first occurred when the Commander activated the portable life support system fan but could not verify flow because the oxygen hoses had inadvertently been left disconnected from the suit. The second delay occurred when both crewmen had inoperable headset microphones because the push-to-talk switch on the remote control unit had not been moved from "off" to "main."

One unusual event occurred prior to turning on the portable life support system oxygen during preparation for the first extravehicular activity. The portable life support system had been connected to the suit, with helmet and gloves on and the fan running. After several minutes in this condition, the suits began to squeeze the crewmen, since they were using up the oxygen by normal breathing. The condition was corrected by turning on the portable life support system oxygen supply. Procedural changes to the checklist will be made to prevent recurrence of this situation.

**TABLE 8.12-I.- EXTRAVEHICULAR MOBILITY UNIT CONSUMABLES**

| Condition | Commander | | Lunar Module Pilot | |
|---|---|---|---|---|
| | Actual | Predicted | Actual | Predicted |
| First extravehicular activity | | | | |
| Time, min | 231 | 210 | 231 | 210 |
| Oxygen, lb | | | | |
| Loaded | 1.254 | 1.27 | 1.266 | 1.27 |
| Consumed | 0.725 | 0.873 | 0.725 | 0.873 |
| Remaining | 0.529 | 0.397 | 0.541 | 0.421 |
| Feedwater, lb | | | | |
| Loaded | 8.56 | 8.60 | 8.50 | 8.60 |
| consumed | 4.75 | 5.4 | 4.69[a] | 5.2 |
| Remaining | 3.81 | 3.2 | 3.8[a] | 3.4 |
| Power, W-h | | | | |
| Initial charge | 282 | 270 | 282 | 270 |
| Consumed | 187 | 130 | 188 | 130 |
| Remaining | 095 | 140 | 94 | 140 |
| Second extravehicular activity | | | | |
| Time, min | 226 | 210 | 222 | 210 |
| Oxygen, lb | | | | |
| Loaded | 1.150 | 1.169 | 1.150 | 1.169 |
| Consumed | 0.695 | 0.886 | 0.720 | 0.849 |
| Remaining | 0.455 | 0.283 | 0.430 | 0.32 |
| Feedwater, lb | | | | |
| Loaded | 8.56 | 8.6 | 8.50 | 8.6 |
| Consumed | 3.89 | 6.2 | 4.69 | 5.8 |
| Remaining | 4.67 | 2.4 | 3.81 | 2.8 |
| Power, W-h | | | | |
| Initial charge | 282 | 270 | 282 | 270 |
| Consumed | 177 | 130 | 177 | 130 |
| Remaining | 105 | 140 | 105 | 140 |

[a] These numbers are factored to include an estimated 1.2 pounds of water lost when the lunar module hatch was accidentally closed, causing the Lunar Module Pilot's portable life support system sublimator to break through.

While the Lunar Module Pilot was in the lunar module prior to the first egress, a loss of feedwater pressure in the portable life support system continued for several minutes. It was found that the lunar module hatch had closed, causing the cabin pressure to increase, which then resulted in a breaking through of the sublimator on the portable life support system. This resulted in a loss of feedwater but did not constrain the extravehicular activity. A procedural change will require that the cabin dump valve remain in the open position.

The portable life support system recharge in preparation for the second extravehicular activity was performed in accordance with established procedures, and the crewmen encountered no significant problems through the completion of the second extravehicular activity.

During the last hookup of the suits to the electronic control assembly prior to ascent, the lunar dust on the wrist locks and suit hose locks caused difficulty in completing these connections. In addition, much dust was carried into the lunar module after the extravehicular periods. Dust may have contaminated certain suit fittings, since during the last suit pressure decay check, both crewmen reported a higher-than-normal suit pressure decay. However, no significant difference in oxygen consumption between the two extravehicular periods was apparent.

The pressure suits operated well throughout the extended use period. The outer protective layer was worn through in the areas where the boots interface with the suit. The Kapton insulation material just below the outer layer also showed wear in these areas. In addition, a minute hole was worn in one of the boot bladders of the Commander's suit. Suit performance was not compromised by this wear, as shown in the following table:

| | Leakage, scc/min | |
|---|---|---|
| | Preflight | Postflight |
| Commander's suit | 105 | 400 |
| Lunar Module Pilot's suit | 51 | 45 |
| Specification value | 180 | 740 |

Note: The leak through the hole in the Commander's boot is estimated to have been about 325 scc/min.

Because the Commander's pressure garment assembly was too short in the legs, considerable discomfort was experienced while wearing the garment in the unpressurized configuration. This misfit resulted from insufficient time in the suit prior to flight to determine the proper adjustment following a last-minute factory rework to correct a leaking boot. Prior to the second extravehicular period, the Lunar Module Pilot corrected a similar condition in his suit by adjusting the laces to lengthen the pressure suit legs.

Twice during the second extravehicular period the Lunar Module Pilot felt a pressure pulse in his suit. A review of data, however, shows no pulse, and this problem is discussed in section 14.3.8.

The performance of the lunar extravehicular visor assembly, which was fitted with side blinders, was excellent. Because the sun angle was very low (near 6 degrees) during extravehicular activities, an additional blinder located at the top center of the visor would have improved visibility. The crewman reduced glare in this situation by blocking out the sun with his hand. An adjustable center blinder, which may be pulled down, will be available for future missions.

The crewmen reported that because of the drying effect of the oxygen atmosphere, it would be desirable to have at least one drink of water during a 4-hour extravehicular period (discussed in section 9.10.3). Future missions will have this capability provided by an in-the-suit drinking bag.

In summary, the calculated metabolic rates of both crewmen during the extravehicular periods were lower than predicted. The extravehicular mobility unit exhibited no significant malfunctions and performed well before and during the extravehicular portions of the mission.

## 8.13 CONSUMABLES

On the Apollo 12 mission, the actual usage of only one consumable for the lunar module deviated by as much as 10 percent from the preflight predicted amount. This consumable was the descent stage batteries. The actual ascent stage water usage was less than predicted because the power load during ascent was less than predicted.

All predicted values in the following tables were calculated before flight.

### 8.13.1 Descent Propulsion System Propellant

The quantities of descent propulsion system propellant loading in the following table were calculated from readings and measured densities prior to lift-off.

| Condition | Actual value, lb | | | Predicted value, lb |
|---|---|---|---|---|
| | Fuel | Oxidizer | Total | |
| Loaded | 7079 | 11 350 | 18 429 | 18 429 |
| Consumed | 6658 | 10 596 | 17 254 | 17 762[a] |
| Remaining at engine cutoff | | | | |
| Tanks | 386 | 693 | | |
| Manifold | 35 | 61 | | |
| Total | 421 | 754 | 1175 | 667 |

[a] Includes allowances for dispersions and contingencies

### 8.13.2 Ascent Propulsion System Propellant

The actual ascent propulsion system propellant usage was within 5 percent of preflight predictions. The loadings in the following table were determined from measured densities prior to lift-off and from weights of off-loaded propellants. A portion of the propellants was used by the reaction control system during ascent stage operations.

| Condition | Actual value, lb | | | Predicted value, lb |
|---|---|---|---|---|
| | Fuel | Oxidizer | Total | |
| Loaded | 2012 | 3224 | 5236 | 5236 |
| Consumed | | | | |
| By ascent Propulsion system | 1831 | 2943 | 4884 | |
| By reaction control system | 31 | 62 | | |
| Total | 1862 | 3005 | 4867 | 4884 |
| Remaining at ascent stage impact | 150 | 219 | 369 | 352 |

### 8.13.3 Reaction Control System Propellant

The preflight planned usage includes 105 pounds for a landing site redesignation maneuver of 60 ft/sec and 2 minutes flying time from 500 feet altitude. The reaction control propellant consumption was calculated from telemetered helium tank pressure histories using the relationships between pressure, volume, and temperature.

| Condition | Actual value, lb | | | Predicted value, lb |
|---|---|---|---|---|
| | Fuel | Oxidizer | Total | |
| Loaded | | | | |
| System A | 108 | 209 | | |
| System B | 108 | 209 | | |
| Total | 216 | 418 | 634 | 633 |
| Consumed to: | | | | |
| Docking | 315 | 305 | | |
| Impact [a] | 433 | 424 | | |
| Remaining at lunar module impact | 201 | 209 | | |

[a] Essentially includes that consumed in the deorbit maneuver.

### 8.13.4 Oxygen

The deviations of actual usage from the predicted consumption result mainly from incomplete telemetry data. When the oxygen is loaded, the pressure and temperature of the oxygen are monitored. In flight, oxygen pressure is the only parameter monitored, and any deviation in temperature causes a change in pressure. Therefore, unrecorded temperature changes can create significant errors in the calculated oxygen consumption. The oxygen used for metabolic purposes is unreasonably low and indicates that temperature changes took place which lend uncertainty to the true indication of actual oxygen usage.

| Condition | Actual value, lb | Predicted value, lb |
|---|---|---|
| Loaded (at lift-off) | | |
| Descent stage | 48.0 | 48.0 |
| Ascent stage | | |
| Tank 1 | 2.4 | 2.4 |
| Tank 2 | 2.4 | 2.4 |
| Total | 4.8 | 4.8 |
| Consumed | | |
| Descent stage | 25.0 | 32.0 |
| Ascent stage | | |
| Tank 1 | 0.6 | |
| Tank 2 | 0 | |
| Total | 0.6 | 1.0 |
| Remaining in descent stage at lunar lift-off | 23.0 | 16.0 |
| Remaining at docking | | |
| Tank 1 | 1.8 | 1.4 |
| Tank 2 | 2.4 | 2.4 |
| Total | 4.2 | 3.8 |

### 8.13.5 Water

The actual water usage was within 13 percent of the preflight predictions. In the following table, the actual quantities loaded and consumed are based on telemetered data. The deviation in the actual usage of ascent-stage water from predicted usage occurred because the do electrical load was lower than predicted.

| Condition | Actual value, lb | Predicted va l ue, lb |
|---|---|---|
| Loaded (at lift-off) | | |
| Descent stage | 252.0 | 250.0 |
| Ascent stage | | |
| Tank 1 | 42.5 | 42.5 |
| Tank 2 | 42.5 | 42.5 |
| Total | 85.0 | 85.0 |
| Consumed | | |
| Descent stage | 169.2 | 174.3 |
| Ascent stage | | |
| Docking | | |
| Tank 1 | 11.2 | 13.5 |
| Tank 2 | 10.5 | 13.5 |
| Total | 21.7 | 27.0 |
| Impact | | |
| Tank 1 | 20.5 | 22.7 |
| Tank 2 | 19.5 | 22.7 |
| Total | 40.0 | 45.4 |
| Remaining in descent stage at lunar lift-off | 82.8 | 75.7 |
| Remaining at ascent stage impact | | |
| Tank 1 | 22 | 19.8 |
| Tank 2 | 23 | 19.8 |
| Total | 45 | 39.6 |

### 8.13.6 Helium

The consumed quantities of helium for the main propulsion systems were in close agreement with predicted amounts. Helium was stored ambiently in the ascent stage and supercritically in the descent stage. Helium loading was nominal, and the usage quantities in the following table were calculated from telemetered data. An additional 1 pound was stored ambiently in the descent stage for valve actuation and is not reflected in the values reported.

| Condition | Descent propulsion | | Ascent propulsion | |
|---|---|---|---|---|
| | Actual value, lb | Predicted value, lb | Actual value, lb | Predicted value, lb |
| Loaded | 48.1 | 48.1 | 13.2 | 13.2 |
| Consumed | 40.1 | 40.1 | 9.2 | 9.2 |
| Remaining | [a]8.0 | 8.0 | [b]4.0 | 4.0 |

[a]At lunar landing. [b] At ascent stage impact.

### 8.13.7 Electrical Power

The crew did not use the interior floodlights according to the checklist, which called for the lights to be at full brightness for all lunar module operations except during the extravehicular and sleep periods. Descent battery usage predicted for these lights was 91 A-h, or 9 percent of the total budget. The lights were used only part of the time during descent and very little while on the surface.

For Apollo 13, predictions will be adjusted to reflect a more practical floodlight operating cycle.

| Batteries | Electrical power consumed, A-h | |
|---|---|---|
| | Actual | Predicted |
| Descent | 1023 | 1147 |
| Ascent (at docking) | [a]230 | 245 |

[a]The failure of the tracking light 1 1/2 hours after lunar lift-off resulted in a saving of 16 A-h.

Commander Charles Conrad, Jr., Commander Module Pilot Richard F. Gordon, and Lunar Module Pilot Alan L. Bean

# 9.0 PILOTS' REPORT

The Apollo 12 mission was similar in most respects to Apollo 11, and this section highlights only those aspects, from the pilots standpoint, which were significantly different from previous flights. In addition, the flight plan was followed very closely. The actual sequence of flight activities was nearly identical to the preflight plan. Figure 9-1 is located at the end of the section for clarity.

## 9.1 TRAINING

The training plan was completed on November 1, 1969, as scheduled. After that date, the training activities were intended as refreshers, except for the detailed planning for the geology traverse scheduled for the second extravehicular excursion. The training time expended provided adequate preparation except in the minor areas to be noted later. Prior to the Apollo 12 preparation, the crew had completed a 1-year training period as the backup crew for Apollo 9, and each pilot was well versed in his particular systems area.

## 9.2 LAUNCH

The countdown progressed normally and ran approximately 20 minutes ahead of schedule after crew ingress. Two system discrepancies were noted during the countdown. A random low-light-level flashing of all "8's" was evident on the display keyboard, and a flashing tuning fork was indicated from the mission event timer on the main display console (section 14.1.1). This keyboard behavior had been experienced before in ground tests and was not considered a significant problem. The central timing equipment was determined to be operating correctly, and the timing problem was isolated to the mission timer, which was not considered essential for launch.

Engine ignition and lift-off were exactly as reported by previous crews. The noise level was such that no earpieces or tubes from the earphones were required. Communications, including the "tower clear" call, were excellent. A potential discharge through the space vehicle was experienced at 36 seconds after lift-off and was noted by the Commander as an illumination of the gray sky through the rendezvous window, as well as an audible and physical sensing of slight transients in the launch vehicle. The master alarm came on immediately, and the following caution lights were illuminated (section 14.1.3): fuel cells 1, 2, and 3; fuel cell disconnect; main bus A and B undervoltage; ac bus 1; and ac bus 1 and 2 overloads. At approximately 50 seconds, the master alarm came on again, indicating an inertial subsystem warning light. Because the attitude reference display at the Commander's station was noted to be rotating, it was concluded that the platform had lost reference because of a low voltage condition. Although the space vehicle at this time had experienced a second potential discharge, the crew was not aware of its occurrence.

The Lunar Module Pilot determined that power was present on both ac buses and had read 24 volts on both main dc buses. Although main bus voltages were low, the decision was made to complete the staging sequence before resetting the fuel cells to allow further troubleshooting by the crew and flight controllers on the ground. It was determined that no short existed, and the ground recommended that the fuel cells be reset. All electrical system warning lights were then reset when the fuel cells were placed back on line. The remainder of powered flight, through orbit insertion, was normal. The stabilization and control system maintained a correct backup inertial reference and would have been adequate for any required abort mode.

One item noted prior to lift-off and at tower jettison was water on spacecraft windows 1, 2, and 3 beneath the boost protective cover. At the time of tower jettison, water had already frozen and later a white powdery deposit became apparent after the frozen water sublimated. These windows remained coated with the deposit throughout the flight, and this condition prevented the best quality photography.

## 9.3 EARTH ORBIT

Because of the potential discharges experienced during launch, several additional checks were performed in earth orbit prior to commitment for translunar injection. These checks included a computer self-check, an E-memory dump, and a verification of thrust vector control. In addition, since platform reference had been lost during launch, a platform alignment and two realignments, to check gyro drift, were conducted. The platform alignment caused the only difficulty when the lack of good dark adaptation made finding stars in the telescope quite difficult. A second factor was that the particular section of the celestial sphere observable at the time was one in which there were no bright stars. The onboard star charts, together with a valid launch reference matrix in the computer, helped appreciably and permitted use of indicated attitudes to locate stars. The stars Rigel and Sirius were used for the platform orientation. Once the platform was aligned, the navigation sightings using auto optics were no problem.

## 9.4 TRANSLUNAR INJECTION

The translunar injection checklist was accomplished as planned and on schedule. The additional checks and alignments provided no appreciable interference, since the timeline was flexible and had been designed to handle such contingencies. The computer program that was loaded into the erasable memory to count down to the launch-vehicle start sequence for translunar injection was a useful addition to onboard procedures. The S-IVB performed all maneuvers, and the translunar injection firing was exactly as planned. The onboard monitoring procedures were excellent and appeared to be adequate for a manual takeover if required.

## 9.5 TRANSLUNAR FLIGHT

### 9.5.1 Transposition and Docking

Physical separation prior to transposition and docking was commenced normally at 3:18:00, but it was observed that the quad-A secondary-fuel and one of the quad-B helium talkbacks indicated barberpole. They were reset immediately with no problems. The only system discrepancy encountered during transposition and docking involved the use of the entry monitor system for measuring the separation velocity provided by the reaction control system. Procedurally, forward thrust was to be applied until the entry monitor system counter indicated minus 100.8 ft/sec. Upon observing the counter shortly after separation, it indicated minus 98 ft/sec; therefore, an accurate measurement of velocity change could not be obtained and forward thrust was continued until separation was assured. The remainder of transposition and docking was conducted in accordance with the checklist. Instead of using the velocity counter to determine separation velocity, the reaction control thrusting should be based on a fixed interval of time. The docking maneuver was performed using autopilot control with 0.5-deg/sec rates and 0.5-degree attitude deadbands. Closing velocities at contact were low and consistent with previous flights.

All post-docking tasks were conducted in accordance with the checklist. Spacecraft ejection was conducted at 04:13:00 and was normal in all respects. The high reaction control propellant consumption encountered with the heavy spacecraft (that is, with the lunar module attached) can be avoided by performing maneuvers using only a 0.2-deg/sec maneuver rate. Also after clearance from the S-IVB is verified, no additional tracking of the S-IVB is needed.

### 9.5.2 Translunar Coast

Activities during translunar coast were similar to those of previous lunar missions and were conducted as planned. The only change from nominal procedures was an early entry into the lunar module to verify that the systems had suffered no damage as a result of the potential discharges during launch. Navigation sightings using the earth limb showed a significant variation in the height of the atmosphere. Future crews should use the apparent visible horizon, instead of the airglow layer, for consistently accurate sightings. Attitude stability was excellent during passive thermal control, which was initiated as planned.

Midcourse Correction

The only midcourse correction required was performed at the second option point with the service propulsion system. This maneuver, the only major change from Apollo 11 during this phase, placed the spacecraft on a "hybrid" non-free-return trajectory (section 5.0). Longitudinal velocity residuals were trimmed to within 0.1 ft/sec.

## 9.6 LUNAR ORBIT INSERTION

The lunar orbit insertion and circularization maneuvers were conducted in accordance with established procedures using the service propulsion system and primary guidance. Residuals were within 0.1 ft/sec about all axes. The computer indicated that the spacecraft was inserted into a 170.0- by 61.8-mile orbit. The planned firing time calculated from ground tracking was 5 minutes 58 seconds, whereas the firing time as observed onboard, was 5 minutes 52 seconds. The circularization maneuver two revolutions later inserted the spacecraft into a 66.3- by 54.7-mile orbit, which included a planned navigation bias as was used in Apollo 11.

### 9.7 LUNAR MODULE CHECKOUT

Activities after circularization were generally routine in nature and closely followed the flight plan. The Commander and the Lunar Module Pilot entered the lunar module for inspection, cleanup, and stowage. During this time, a scheduled landmark tracking of a crater (designated H-1) in the vicinity of Fra Mauro was normal in all respects and established procedures were used without difficulty.

Lunar module checkout prior to descent orbit insertion was commenced on time after completion of suiting and proceeded normally. Two new procedures were used during this flight to eliminate unnecessary orbital perturbations so that state vectors for descent orbital insertion would be known accurately. All docked maneuvers were conducted using balanced thrust coupling, and the soft undocking was performed in a radial attitude. The soft undocking was normal in all respects and procedurally similar to that for Apollo 9. The first separation maneuver was accomplished by firing the service module reaction control thrusters in the plus-2 direction while in a local horizontal attitude.

Lunar module power-up varied in two aspects from planned procedures. The crew had decided to evaluate in real time the suit donning in the command module and, if practical, to suit the Lunar Module Pilot and then the Commander prior to initial transfer. This procedure was shown to be feasible, and the Lunar Module Pilot was fully suited when he entered the lunar module for power-up. During preflight simulations of power-up, it was apparent that several scheduled events in the pre-descent timeline had a minimal time allotted because of the scheduled landmark tracking and platform alignment prior to reaction control system checks, which required network coverage. Therefore, procedures were established with the ground to gain additional time for possible contingencies and to perform the reaction control hot- and cold-fire checks that could be done prior to landmark tracking. All systems checked out well on initial power-up, and as a result, the timeline in the lunar module remained about 40 minutes ahead of schedule after the first revolution. Undocking occurred on time, with the only unexpected events being an 1106 alarm upon computer power-up, the validity of rendezvous radar self-test values in the checklist, and a low rendezvous radar transmitter power output.

### 9.8 DESCENT ORBIT INSERTION

The lunar module was pitched and yawed at undocking to the planned inertial attitude, and then a yaw maneuver was manually initiated to achieve the proper attitude for automatic sighting maneuvers. Three automatic maneuvers were performed, two for star sightings and one for the landing-point-designator calibration. A maneuver was then completed to the descent orbit insertion attitude, which was maintained until after ignition. The descent orbit insertion maneuver was initiated on time and velocity residuals, as indicated by the primary system, were very low and in close agreement with those displayed by the abort guidance system. Therefore, no velocity trimming was necessary. Soon after descent orbit insertion, the lunar module was maneuvered to the attitude for powered descent initiation. Throughout the flight phase from undocking to powered descent, maneuvering was held to a minimum so as not to perturb the established orbit.

### 9.9 POWERED DESCENT

The powered descent initiation program was selected twice in the timeline; the first was to permit a quick look at system operation about 25 minutes after descent orbit insertion and the second was 8 minutes prior to powered descent initiation after receiving the latest network update. Powered descent initiation and throttle-up were on time. Throughout the major portion of descent, considerable reaction control thruster activity, which has been attributed to fuel slosh (see section 4.2.2) was noted. The landing point update was received and entered at approximately 1-1/2 minutes after powered descent initiation. The landing radar altitude and velocity lights went out, indicating proper radar acquisition, approximately 4 seconds apart at altitudes near 41 000 feet.

Throttle-down occurred within 1 second of the predicted time. The abort guidance system readouts remained consistent with the primary system at all times, and the abort guidance altitude was updated three times during descent. Computer switchover to the landing program occurred on time. Immediately after pitchover, lunar surface features seen through the window were not recognizable. The field of view and the lunar surface detail are greater than in the simulator, and training photographs are not adequate preparation for the first look out the window. However, with the first sighting through the landing point designator at the nominal 42-degree angle, all the planned landmarks became very obvious. The subsequent landing-point-designator angles indicated a zero crossrange error and a downrange error that was either very small or non-existent. Therefore, no early landing-site redesignations were required.

The first redesignation, a 2-degree right correction, was made late in the descent to maneuver out of the center of the Surveyor crater. Several redesignations were then made, both long and short (fig. 4-11), according to the condition apparent at the time. The preselected landing site at the 4-o'clock position (from north) around Surveyor crater did not appear to be suitable upon reaching an altitude of 800 feet, and a more suitable site appeared to be one near the 2-o'clock position. The manual descent program was entered at approximately 400 feet altitude to prevent an apparent downrange miss and to maneuver to the left. A steeper-than-normal descent was made into the final landing site. Dust was first noted at approximately 175 feet in altitude. The approach angle was approximately 40 degrees to the surface slope. A left translation was easily initiated and subsequently stopped to maneuver over to the landing site. The last 100 feet were made at a descent rate of approximately 2 ft/sec. Prior to that time during the landing phase, the maximum descent rate was 6 ft/sec. The dust continued to build up until the ground was completely obscured during approximately the last 50 feet of descent (section 6.1). Although the cross-pointer velocity indicator was not checked prior to 50 feet, at which point ground reference was obscured, the indicator read zero, indicating zero crossrange and downrange velocities. All quoted altitudes during final descent were based on computer values, as read by the Lunar Module Pilot, and the computer indicated 19 feet in altitude after touchdown. The computer altitude indication is referenced to landing-site radius and ideally should have been approximately 4 feet.

Although the lateral velocities were actually zero, as indicated, a possible indicator failure was suspected, and control was continued half visual and half by instruments. The Commander was scanning the instruments when the lunar contact light illuminated. The engine was subsequently shut down. The touchdown which followed was very gentle, and during extravehicular activity, a postflight examination of the gear struts and pads indicated zero translation and very low sink rates at touchdown.

The descent fuel and oxidizer tanks were vented as planned, and the "stay" decisions were received on time. Two lunar surface alignments were performed, and the lunar module was then powered down to the configuration for extravehicular preparation.

## 9.10 LUNAR SURFACE ACTIVITY

### 9.10.1 Preparation for Initial Egress

Initial egress to the surface occurred later than planned, because more time than anticipated was spent in locating the lunar module position on the surface prior to egress. It also took longer than expected to configure the suit hoses and position communication switches from memory, instead of a specific checklist callout. The checklist was accurate and adequate for preparing all equipment for extravehicular activity. The one-g high fidelity preflight simulation of preparation for extravehicular activity was extremely beneficial and resulted in both crewmen preparing for surface activity in a rather routine fashion.

Defining the exact location of the lunar module proved to be difficult because of the limited field of view through the windows, the general tendency to underestimate distances (sometimes as much as 100 percent), and the difficulty in seeing even large craters outside a distance of several hundred feet. An accurate position of the spacecraft was easily determined after egress to the lunar surface.

Communications while using the backpack equipment within the cabin were excellent at all times, and no garbling with the antenna either stowed or deployed was experienced. The improved circuit breaker guards were effective in that no circuit breakers were accidentally opened or closed throughout lunar module activities.

During the 4- or 5-minute period immediately after donning the helmet and gloves, but prior to the integrity check of the extravehicular mobility unit, the suits tended to shrink around both crewmen and resulted in a rather uncomfortable condition. This problem was solved by momentarily actuating the oxygen valve to place about 0.5 psi in the suit.

Cabin depressurization without the filter installed on the dump valve did not take excessive time. It was possible to "peel open" the forward hatch from the upper left-hand corner at a cabin pressure slightly higher than that associated with use of the hatch handle only. It took about 5 seconds after the corner of the hatch was peeled open before the cabin pressure lowered sufficiently for the hatch to swing to the full-open position.

### 9.10.2 Egress

Egress and ingress were found to be relatively simple and similar to preflight simulations. On the first egress, a 6-inch tear was made in the outside thermal skin of the door by contact with the lower left-hand corner of the backpack because the egressing crewman was slightly misaligned to the left of the hatch centerline. Despite this occurrence, the size and shape of the hatch are considered to be completely adequate.

After the Commander had first egressed to the surface, the Lunar Module Pilot moved back and forth across the cockpit to photograph the Commander and to receive transferred equipment. During this time, the hatch was inadvertently swung near the closed position, and outgassing from the portable life support system sublimator provided enough pressure to close the hatch. The cabin pressure then rose slightly and caused a water breakthrough of the sublimator, with associated caution-and-warning alarms. When the cause of the breakthrough was discovered, full operation of the sublimator was quickly restored by opening the hatch and returning the partially pressurized cabin to a full vacuum.

After the Lunar Module Pilot had egressed (fig. 9-2), he had difficulty in closing the door from the full-open to a partial position, since there is no exterior handle provided. The flap that covers the hatch lock handle cannot be reached from outside the spacecraft with the door full open, and the only other protuberance, the door covering the dump valve, is so close to the hinge line that considerable force must be used to close the door.

NASA-S-70-589

Figure 9-2 Lunar Module Pilot descending to the lunar surface.

Although neither crewman noted a tendency for his boots to slip on the surface, mobility and stability were generally as reported in Apollo 11. Acclimation took less than 5 minutes and permitted each crewman to begin the nominal timeline immediately. The 1/6g and the partial gravity simulators were excellent training devices for learning the most efficient ways to move about on the lunar surface. The 5-minute familiarization period at the beginning of each extravehicular period is ideal.

Figure 9-3 Lunar Module Pilot extracting the fuel cask. The radioisotope thermoelectric generator is shown near the crewman.

### 9.10.3 Extravehicular Mobility Unit Operation

The performance of the extravehicular mobility unit was faultless. Although the maximum cooling position of the portable life support system diverter valve had been used frequently during preflight testing involving high workloads, the minimum cooling position with occasional 1-minute intermediate cooling selection was completely adequate to perform even the most strenuous lunar surface work. Continued use of the minimum cooling configuration was surprising, since both crewmembers felt that they were working at about the maximum practical level needed for lunar surface activity. Even at these workloads, it was believed that extravehicular periods could be extended to as many as 8 hours without excessive tiring. During the two 4-hour work periods for this flight, it would have been desirable to have at least one drink of water because of the drying effect of the oxygen atmosphere. Extravehicular periods of longer duration will require some water and possibly energy in the form of liquid food. Although the suit was completely adequate to accomplish mission objectives, the efficiency of the overall lunar surface work could be enhanced by 20 or 30 percent if it were possible to bend over and retrieve samples from the surface. [Ed. note: A suit with this capability is planned for Apollo 16.]

Although the gloves were found to be clumsy for changing camera magazines, they were completely acceptable for all other tasks. The Lunar Module Pilot felt a slight heat soak-through in the palms of the gloves when he carried the lunar tools or gripped the hammer, such as when pounding in a core tube.

The checklist on the glove cuff was an excellent device and provided good readability and ample space for information without interfering with normal tasks.

It was difficult to walk "heel-toe, heel-toe" on the lunar surface in a fashion similar to an earth walk because of suit mobility restriction. As reported by the Apollo 11 crew, it was much easier to lope about in a stiff-legged, flat-foot fashion. Because of the reduced gravity, there is a brief period when both feet are off the ground, a condition which gives the crewman the impression he is moving rapidly. However, as simulated with the centrifuge partial gravity simulator before flight, the surface movement was only about 4 ft/sec, a normal earth walking pace.

### 9.10.4 Extravehicular Visibility

Lunar surface visibility was not too unlike earth visibility, except that the sun was extremely bright and there was a pronounced color effect on both the rocks and soil. Cross-sun and down-sun viewing was not hindered to any great degree. When viewing up sun, it was necessary to use a hand to shield the eyes, because the usual technique of "squinting" the eyes did not sufficiently eliminate the bright solar glare. It would have been helpful to have an opaque upper visor on the helmet similar to the two side visors provided for this flight. It was difficult to view down sun exactly along the zero-phase direction. This deficiency did not hinder normal lunar surface operations because the eyes could be scanned back and forth across this bright zone for visual assimilation. Objects in shadows could be seen with only a slight amount of dark adaptation. The apparent color of the lunar surface depended on both the angle of sun incidence and the angle of viewing. At the low sun angles during the first extravehicular period, both the soil and the rocks exhibited a slight gray color. On the second extravehicular excursion, the same rocks and soil appeared to be more a light brown color. Because the sun angle had such a pronounced effect on color, minerals within the rocks were difficult to identify, even when the rocks were held in the hand and under the best possible lighting. During the first extravehicular period, the slope at the Surveyor location was in shadow, and this slope appeared to have an inclination of about 35 degrees. However, the next day after the sun had risen sufficiently to place the Surveyor slope in sunlight, the inclination appeared to be 10 or 15 degrees, which is closer to the true value.

### 9.10.5 Lunar Surface Experiments

The deployment handle for the door to the modularized equipment stowage assembly in the descent stage could not be pulled from its socket. Therefore, the door was lowered by pulling on the cable extending from the handle to the release mechanism. The experiments package was then easily unloaded. The booms should be eliminated since there is no pronounced tendency to be unbalanced when removing the large experiment packages from the lunar module. The straps which open the scientific equipment bay doors, extend the booms, and lower the packages and fuel cask were excessive in length. Considerable effort was required to keep them from tangling. A smoother and faster unloading could have been accomplished if the straps had been considerably shorter and if a manual unloading technique had been used. The fuel cask guard (part of the experiment equipment) was also not needed.

The fuel element stuck in the cask (fig. 9-3) and could not be removed with normal force. By striking the side of the cask with a hammer and exerting a positive pull on the element, it was possible to extend the element an additional 1/8 inch or so for each hammer blow. After the element had been extended about an inch, it became free and was removed and placed in the radioisotope thermal generator. The thermal generator was easy to fuel. Heat radiating from the fuel element was noticeable through the gloves and during the walk to the deployment site but was never objectionable.

The experiment packages were deployed to a distance of about 425 feet. The necessity for gripping the carry bar tightly was tiring to the hands. Some type of over-the-neck strap would probably be advantageous for deployment distances beyond 300 feet. Selection of a suitable deployment site was not difficult in the Apollo 12 landing area. The central station deployed normally. Leveling and aligning of the antenna were performed according to the checklist.

Special care had to be taken when deploying the power cable, since the bracket had been heated by the thermal generator. This deployment was necessarily a two-man operation. The silver and black decals on the equipment were very difficult to read in the bright sunlight. After the power plug was connected to the central station, the shorting-plug current could not be read because the needle was not visible in the instrument window. It is possible that the shorting plug had already been depressed prior to the intended time.

The passive seismic experiment was difficult to deploy because the mounting stool did not provide sufficient protection against inadvertent contact of the bottom of the experiment with the lunar surface. To overcome this deficiency, it was first necessary for the crewman to dig a small hole with his boot, a procedure which was time consuming and not very precise. The thermal skirt would not lie flat when fully deployed, and it was necessary to use Boyd bolts and clumps of lunar surface material to hold the skirt down. Leveling the experiment was simple using the bubble; however, the metal ball leveling device was useless because of the lack of adequate damping of ball motion.

Deployment of the suprathermal ion detector was difficult because of the short distance between the three legs. The ground screen on which the detector was to sit had a spring loaded over-center feature which made it difficult to deploy. The protective lid, designed to be released by ground command, opened accidentally three times during deployment and had to be reclosed. The deployment operation was therefore time consuming, and the cover was left open the last time, since the experiment was already in place.

The cold cathode gage could not be deployed with the aperture facing west because the power cable was too stiff. Once the gage was set in the proper position, the cable would move it to an aperture-down attitude. After about 10 attempts, which required both crewmen, the gage accidentally assumed an aperture-up position and was left in this attitude since it appeared to function normally.

It was impossible to work with the various pieces of experiment equipment without getting them dusty. Dust got on all experiments during off-loading, transporting, and deployment, both as a result of the equipment physically touching the lunar surface and from dust particles scattered by the crewmen's boots during the deployment operation. Because there does not appear to be a simple means of alleviating this dust condition, it should become a design condition. Although both experiment package tools worked well, the deployment could have been more efficient if the tools had been from 2 to 5 inches longer. The difficulty in fitting and locking both tools in most of the experiment receptacles was frustrating and time consuming. Looser tolerances would probably eliminate the problem.

The environmental sample and the gas sample were easy to collect in the container provided, but there was a noticeable binding of the threads when replacing the screw-on cap. The binding could have been caused by a thermal problem, operation in a vacuum, or the threads being coated with lunar dust. Although the lid was screwed on as tightly as possible, the gas sample did not retain a good vacuum during the trip back to earth.

The solar wind collector was deployed easily but was impossible to roll up. The collector could be rolled up in a rather normal fashion for approximately the first 8 inches, but beyond that point the foil would not easily bend around the roller. The problem was apparently caused by an increase in foil or foil backing tape stiffness, rather than by roller spring torque. The foil was rolled by hand before stowage in the Teflon bag in the sample return container. The Teflon bag was too short and did not permit the foil to be rolled sufficiently to keep dirt within the sample box from getting on the solar wind collector.

### 9.10.6 Surveyor Inspection

The entire Surveyor operation was very smooth. The bag and tools were removed from the descent stage storage compartment and placed on the Commander's back with relative ease. This location did not hinder mobility or stability and should be considered as a location for other bags and tools on future missions.

The Surveyor was sitting on a slope of approximately 12 degrees. All components were covered with a very tenacious dust, not unlike that found on an automobile that has been driven through several mud puddles and allowed to dry. While the dust was on all sides of the Surveyor, it was not uniform around each specific item. Generally, the dust was, thickest on the areas that were most easily viewed when walking around the spacecraft. For example, the side of a tube or strut that faced the interior of the Surveyor was relatively clean when compared to a side facing outward.

Retrieving the television camera was not difficult using the cutting tool. The tubes appeared to sever in a more brittle manner than the new tubes of the same material used in preflight exercises. The electrical cable insulation had aged and appeared to have the texture of old asbestos. The mirrors on the surface of the electronic packages were generally in good condition. A few cracks were seen but no large pittings. The only mirrors that had become unbonded and separated were those on the flight control electronics package. As a bonus, the Surveyor scoop was removed. Although the steel tape was thin enough to bend in the shears and could not be cut, the end attached to the scoop became debonded when the tape was twisted with the cutter. Several rock samples were collected in the field of view of the Surveyor television camera for comparison with original photographs. On the return traverse, the added weight of the Surveyor components and samples on the crewman's back did not appear to affect either stability or mobility.

### 9.10.7 Lunar Surface Tools

The handtool carrier was light but was still troublesome to carry about. When a number of samples had been accumulated, it was tiring to hold the carrier at arm's length so that rapid movement was possible. If a means could be found to attach the carrier to the back of the portable life support system during the traverse from one geology site to another, the total geology operation could be carried out more efficiently. It was generally necessary to set the carrier down with great care to prevent it from tipping over. The practicality of a pushed or towed vehicle for transporting equipment, tools, and samples over the surface could not be resolved from the work performed in this mission. However, certain constraints, such as the dust which would be set in motion by any wheels, must be considered in the design of such a vehicle. Also, under the light gravity, objects carried on such a conveyance would have to be positively restrained.

The hammer proved to be an effective tool. Since arm motion is inaccurate in the pressurized suit, the front end of the hammer was generally not used when driving a core tube because its striking area was too small, and the side of the hammer was more useful. The pick portion of the hammer is of questionable value because of the danger of flying fragments. The thin metallic coating on the hammer fractured and flew off during normal hammering operations.

The tongs are from 3 to 5 inches too short to select samples from the lunar surface easily. Further, their limited jaw size (fig. 9-4) allows selection of only very small rocks. Because of time limitations, the optimum sample size was larger than either the tongs could pick up or the sample bags would hold. The individual documented sample containers and tear-away sample bags were too small to hold the most desirable samples observed, and the tear-away sample bags were the easier of the two types to use. Furthermore, the two holding arms for the documented sample containers became bent because of interference with the suit during normal movement.

The extension handle was also from 3 to 5 inches too short for optimum use with the shovel. The upper collar that mates with the aseptic sampler is no longer required and could be removed. The locking collar for the shovel or core tube was binding slightly by the end of the second excursion, probably because of dust collection in the mechanism. The shovel was used to dig trenches, as well as to collect soil samples. With the present extension handle for the shovel, it was only possible to dig trenches about 8 inches in depth. Trenching operations were very time consuming. Because of the continuous mantle of dust that coats most of the lunar surface, trenching should be deeper and more frequent on future missions. A specific trenching tool should be used.

Single core tubes were easy to drive and did not require augering. Friction would steadily build up as the tube went into the lunar soil. Driving the double core tube required stronger hammer blows. The soil within the core tube compacts somewhat during the driving operation, particularly for a double-core-tube specimen. Therefore, space remains in the tube when it has been driven to its full length.

### 9.10.8 Lunar Surface Equipment

The single-strap lunar surface conveyor (fig. 9-5) was easy to deploy and generally performed satisfactorily. The end of the strap resting on the surface collects dust, which is subsequently deposited on the crewmen and in the lunar module cabin. The metal pin that retains the lunar module end of the conveyor was not large enough to prevent it from slipping out of the yoke. By the end of the second extravehicular period, the lock buttons on the two hooks were extremely difficult to operate because of accumulated dust. This locking feature is not necessary.

Figure 9-4 Lunar sample collection using tongs.

Figure 9-5 Commander operating equipment conveyor.

The contingency sample could be taken more efficiently if the retrieval handle were 4 or 5 inches longer. Actually, the contingency sample turned out to be a fortunate choice, since two of the more unusual rocks collected during the lunar stay were part of this sample.

The Teflon saddle bags tended to retain their folded shape when removed from the sample return containers. After the first extravehicular period, the bags cracked at several points along the crease lines.

Closing of the sample return containers was not difficult and was similar to that experienced during 1/6g simulations in an airplane. The seal for the sample return container lid became coated with considerable dust when the documented samples were being loaded into the container. Although the surface was then cleaned with a brush, the container did not maintain a good vacuum during the return to earth.

The television camera operated properly while still stowed in the descent stage equipment compartment. However, while the camera was being transferred to the deployed surface position, the camera was accidentally pointed at either the sun or the sun's reflection on the descent stage and the vidicon tube apparently burned out (section 143.1). It is believed the camera is satisfactory for lunar surface work but will have to be handled more cautiously. The markings on the lens for focus, zoom, and aperture were difficult to use because of the bright sun and the fact that the camera, when mounted on the tripod is not very close to the crewman's eyes. A television monitor, similar to that used in the command module, would be desirable for lunar surface operations. A flight configuration television camera should be furnished for preflight training and a qualified engineer should be assigned to review crew procedures prior to flight to insure their adequacy. Although the television cable lay flat on the ground, it still provided a severe foot entanglement problem when a crewman was operating near the spacecraft, particularly when near the descent stage equipment compartment. Routing the cable from a descent stage quadrant other than the one on which the storage assembly is located would help.

The erectable antenna was easy to deploy on its tripod but difficult to align. The entire unit tends to move about when the handcrank is used to adjust the antenna dish. The alignment sight does not have a sufficient field of view and must be precisely aligned to contain the earth's image. Since this function is the purpose of the sight, it may be desirable to add an additional sight with a larger field of view. Although one-man deployment was satisfactory, both crewmembers were required to align the antenna.

All shades on the contrast charts could be seen under the conditions tested. One of the charts was accidentally dropped to the surface, and the dust coating rendered it unusable. The other two charts were used to look at the two extreme lighting conditions, up sun and down sun on the walls of a crater.

The exterior of both cameras became extremely dusty on the lunar surface. It is believed that some dirt was on the lens, although this condition was difficult to detect because the lenses were recessed. Cleaning the lens was not possible but would have been desirable. Toward the end of the second extravehicular period, the fluted thumbwheel on the screw that attaches the camera to the camera mounting bracket, which then attaches to the front of the suit, worked free from the screw. The camera could no longer be mounted to the bracket or the suit and was therefore not used for the remainder of the extravehicular phase (see section 14.3.10).

Adequate time was not available to take full advantage of the capability of the lunar surface closeup camera. The camera performed satisfactorily, except that the film counter would not work. An increase of the spring force holding the extension shield down would prevent accidental movement of the camera when taking photographs.

The 30-foot tether was not used because of the ease of operating on the 12 degree slope of Surveyor crater. However, the tether should be retained for future missions, because the crew may attempt to collect samples in craters with steep sides. A 100-foot tether would be ideal for determining whether or not a specific crater wall was adequate for descent.

The annotated geology charts were excellent aids, both in the lunar module and on the lunar surface, for planning the traverse and in locating surface features. The photo map on one side of a chart depicted the traverse, and the other side of the chart contained descriptions of geologically interesting items to investigate. The photo map should be graphically enhanced so that the size and shape of craters and/or hills can be more easily seen. Use of multicolored areas to depict the geological units should be retained, but the colors should be subdued to enhance the ability to read crater size and shape. Although multiple alternate traverses may be planned, only one prime traverse should be detailed for subsequent missions, primarily because a landing within walking distance of the planned traverse is probable. Efficiency on the surface can be further enhanced by performing the actual prime traverse under simulated conditions during preflight training.

### 9.10.9 Activity in the Spacecraft on the Surface

Cabin repressurization after each extravehicular period was positive and rapid. Once inside the spacecraft, the dust on the suits became a significant problem. Considerable dirt had adhered to the boots and gloves and to the lower portions of the suits. There were fillets of dirt around the interior angles of the oxygen hose connectors on the suit. The suit material just beneath the top of the lunar boots chafed sufficiently to wear through the outer suit layer in several spots. The dust and dirt resulted in a very pronounced increase in the operating force necessary to open and close the wrist rings and the oxygen hose connectors. The Commander's suit had no leakage, either prior to launch or prior to the first extravehicular activity. Just before his second egress, the leak rate was 0.15 psi/min and, prior to cabin depressurization for equipment jettison, was 0.25 psi/min. If the suit zippers had been operated for any reason, suit leakage might have exceeded the 0.30 psi/min limit of the integrity check. (Editor's note: See section 8.12)

After ascent orbit insertion, when the spacecraft was again subject to a zero-g environment, a great quantity of dust and small particles floated free within the cabin. This dust made breathing without the helmet difficult and hazardous, and enough dust and particles were present in the cabin atmosphere to affect vision (section 6.2). Some type of throwaway overgarment for use on the lunar surface may be necessary. During the transearth coast phase, it was noticed that much of the dust which had adhered to equipment (such as the camera magazines) while on the lunar surface had floated free in the zero-g condition, leaving the equipment relatively clean. This fact was also true of the suits, since they were not as dusty after flight as they were on the surface after final ingress.

The sleeping hammocks were particularly good under the reduced gravity conditions. The noise within the lunar module was loud, but not enough to prevent adequate sleep, and the earplugs were not used. The only noise problem was caused by the coolant pump changing frequency several times during the night. Temperature control was satisfactory during the sleep period, and the liquid cooling garment pump was not used. The suit hoses were generally disconnected from the suit, with the suit isolation valves open. The hoses were connected to the suit only a few times, as necessary to cool the feet and lower legs.

When the Commander connected his suit hoses after the first extravehicular activity, he felt free water in his suit. Upon removing the inlet hose, two or three 1/2-inch globules of water were blown from the system. Although both fans and both water separators were operated in an attempt to eliminate the problem, the presence of free water in the Commander's suit loop occurred subsequent to each cabin repressurization and provided a mildly uncomfortable environment. The Lunar Module Pilot's hoses provided adequately dry air at all times.

Recharging of the portable life support system with oxygen or water was easily accomplished, as was the changing of the lithium hydroxide cartridge and the battery. In both recharges, the oxygen filled to above the 80-percent mark. The scale used for weighing the water remaining in the portable life support system prior to recharge was not satisfactory, since it could not be zeroed under the 1/6g conditions. Section 8.11.2 presents a discussion of this problem.

The storage of the Surveyor bag and its components in the lunar module was completely satisfactory. This area would provide an ideal location for permanent type stowage of loose items returned from the moon. The extra 15 pounds of rocks were lashed just aft of the two oxygen purge systems on the cabin floor.

Cabin depressurization for equipment jettison was routine. Jettisoning of the equipment soft pack is most easily accomplished by leaning over and shoving it out the hatch. The portable life support systems were jettisoned by placing them in front of the hatch, tipping them slightly, and dropkicking them out the hatch. With this technique, all items could safely clear the descent stage.

Lunar surface alignments were performed as a two-man operation. The Commander manually recorded and inserted data into the computer, while the Lunar Module Pilot sighted through the optics, punched the mark button, and read the spiral and cursor angles to the Commander. It was impossible to keep the eye centered on the eyepiece and view stars that were greater than 20 degrees from the center of the field of view. It was also impossible to have both the stars and the reticle in focus with the same setting. For this reason, stars should be selected near the center of the detent. If none of the 37 star locations stored in the erasable memory are suitable for sightings, any of the other 400 Apollo stars available from the ground can be used by entering the half-unit vectors. This substitution is not time consuming and is operationally acceptable. Because the landing site was located at the 23-degree west longitude, visibility out the three forward detents was excellent. Enough stars were visible to easily identify major constellations in these three detent positions. The left-rear detent was streaked somewhat, yet several bright stars were visible. The rear and the right-rear detents were completely washed out by sunlight.

## 9.11 ASCENT, RENDEZVOUS, AND DOCKING

### 9.11.1 Ascent

The first items on the pre-ascent checklist were commenced 2 hours 50 minutes before scheduled lift-off (power-up and lunar surface alignment operations). There were no major deviations from the checklist, and lift-off occurred on time. At lift-off, an abundance of silver- and gold-colored insulation material was noted traveling radially outward parallel to the lunar surface, as reported in Apollo 11. Pitchover was smooth, and the yaw maneuver was performed manually 1 minute after liftoff. The rendezvous program was targeted in real time to give a zero change in velocity for the constant differential height maneuver during rendezvous. The comparison of actual with planned velocity showed a slight increase over nominal values throughout ascent, indicating a slightly higher-than-average engine performance. The Lunar Module Pilot closed the ascent feed valves at 200 ft/sec remaining to shutdown, in accordance with the checklist. However, the left-main shutoff valve indicated it was still closed, and because the Commander's attention was distracted by this problem, he did not place the ascent-engine arm switch to "off" at 100 ft/sec remaining, as planned. The late placement of this switch caused a 30-ft/sec overturn, which was immediately removed with reaction control trimming. The main shutoff valve indicated closed, after recycling of the control, and it was not apparent whether the problem was in the talkback indicator or in the valve itself (section 8.11.1 is a discussion of this problem). The ascent stage could not be tracked by the Command Module Pilot during the insertion firing; therefore, an automatic maneuver was conducted in the command and service module to an attitude compatible with both radar acquisition and sextant tracking.

### 9.11.2 Rendezvous

The post-insertion checklist and inflight alignment in the lunar module were completed on time. The inflight alignment was performed as a two-man operation in a manner similar to the surface alignments. It was easy to adjust the reticle brightness and to focus the optics so that the target star and reticle were of good relative brightness and definition. An important consideration in getting accurate alignments was insuring that the eye was accurately centered in the eyepiece.

The handling characteristics of the lightweight ascent stage in the primary guidance pulse mode were satisfactory for alignments and manual tracking with the rendezvous radar. Rendezvous radar navigation was initiated, and the first update gave only small errors for range and range rate. These values were therefore accepted, and no other out-of-limit dispersions were noted throughout the remainder of the rendezvous. All out-of-plane computations were less than the value which would have necessitated a firing; therefore, no out-of-plane corrections were made prior to terminal phase initiation. The terminal phase initiation solution showed a plus 1.5-ft/sec out-of-plane correction, and this value was combined with the inplane maneuver and executed. The computations showed a constant 17.5-mile height differential throughout rendezvous. All command module and lunar module solutions were in good agreement (table 5-VII).

Although the midcourse corrections were small, both solutions were executed. It was not necessary to make any line-of-sight corrections in the lunar module until at a range of approximately 1000 feet from the command module, and these corrections were very small. The velocity limits for all braking gates were met, with the first gate at 6000 feet range requiring a velocity reduction from 38 to 30 ft/sec. The passive rendezvous procedures for the command module were normal in all respects. The ground uplinked the lunar module state vector immediately after insertion, and a platform alignment was conducted according to the checklist. This procedure was completed ahead of the nominal timeline and permitted orbital navigation to be commenced early. The VHF ranging system broke lock twice in the subsequent tracking timeline. For the out-of-plane solution, nine VHF ranging and 14 optics marks were obtained. The only procedural discrepancy noted was the initial few state-vector solutions did not converge as rapidly as expected; however, a solution for coelliptic sequence initiation of 38.8 ft/sec was eventually obtained. The command module navigation operation was continued, with the final computation completed on time after 14 VHF and 21 optics marks had been obtained. The final command module solutions for coelliptic sequence initiation and the constant differential height maneuver were comparable to those of the lunar module. The rendezvous timeline through the constant differential height maneuver was nominal in all respects.

Although sun shafting was evident in the sextant, eight optics marks were obtained before darkness. When the lunar module went into darkness, the Command Module Pilot observed that the lunar module tracking light was inoperative. All checks on board the lunar module indicated that switches were in the proper configuration, and it was assumed that the tracking light failed subsequent to coelliptic sequence initiation. Therefore, the remainder of the command module rendezvous operations were conducted using VHF ranging only. The solutions for terminal phase initiation in both vehicles were again comparable. As was known prior to flight, both midcourse correction solutions in the command module would be inaccurate when only VHF ranging was used.

### 9.11.3 Docking

The command module digital autopilot was set to narrow deadband and used to perform the pitch and yaw maneuver for the docking operation. At capture latch engagement, the command and service module control mode was then changed to free, while the lunar module remained in attitude-hold, narrow deadband. There were no noticeable docking transients or lunar module reaction control thruster firings. A slight attitude adjustment was made with the command and service module, and the probe was then retracted for a hard dock. Closing rates at contact are estimated to have been about 0.2 or 0.3 ft/sec.

### 9.11.4 Crew and Equipment Transfer and Separation

After docking, the tunnel was cleared, lunar module equipment was transferred to the command module, and command module jettisonable equipment was placed in the lunar module. All activities during this period were completely normal.

The transfer of equipment between both vehicles was impeded by the large amounts of dust and debris in the lunar module. Therefore, the timeline became very tight in meeting the schedule for lunar module jettison. However, the checklist and the flight plan were completed satisfactorily. On future flights, at least an additional half hour should be allowed for this activity. Lunar module jettison and the subsequent command and service module separation maneuver were conducted in accordance with flight plan procedures.

## 9.12 LUNAR ORBIT ACTIVITIES

### 9.12.1 Lunar Module Location

On the first revolution after lunar landing, simultaneous tracking from both spacecraft was conducted to enable the ground to determine the exact location of the landing site. Lunar landmark 193 was tracked from the command module, and the lunar module tracked the command module using the rendezvous radar. On the next pass, the lunar module was tracked from the command module using the latitude and longitude of the landing site as supplied by the ground. The technique involved finding the "snowman" (section 4.3) in the telescope and locating the lunar module through knowledge that the vehicle had landed on the northwest side of the Surveyor crater. The telescope was positioned as close as possible to the landing site, and the sextant was then used to find the lunar module, which appeared as a bright object with a long pencil-thin shadow. Recollections after the flight included the fact that the entire descent stage was observed in the sextant. As the command module passed through the zenith, the Surveyor was observed as a bright spot in the shadow of the Surveyor crater. On the next pass, the 16-mm sequence camera was mounted on the sextant to obtain pictures of the landing site.

In the command module orbital revolution before lift-off, the lunar module could not be acquired in the command module sextant either by using auto-optics, which did not point the sextant axis at the lunar module, or by manually positioning the sextant. The telescope should be used as the searching device, rather than the sextant, which has a much smaller field of view. Once the target area is found in the telescope, sighting can be transferred to the sextant. Just prior to lift-off, a second attempt was made to locate the lunar module, and this time the vehicle was observed in the sextant once the Surveyor crater and associated snowman (section 4.3) were found by means of the telescope.

### 9.12.2 Lunar Orbit Plane Changes

A platform alignment was conducted in the command module to prepare for the first out-of-plane maneuver. The techniques employed by the Command Module Pilot to make this maneuver unassisted made maximum use of ground monitoring and assistance. The first lunar orbit plane change was an 18-second service propulsion maneuver, which was nominal and required no velocity trimming. At the completion of this firing, an additional alignment was conducted to the landing-site orientation. The second lunar orbit plane change was conducted, using the service propulsion engine under primary guidance and control, to provide better orbit coverage for the bootstrap photography, described later. This maneuver was normal in all respects, with the exception of a slight tendency for the vehicle to exhibit a "dutch roll" during the maneuver (section 7.6). However, guidance during the maneuver appeared to be normal, and no action was taken. Velocity residuals were low, and no trimming was required.

### 9.12.3 Multispectral Photography

The multispectral photography experiment was conducted from the command module while the lunar module was on the surface and was excellent from an operational viewpoint. No difficulties were encountered in camera assembly or installation on the hatch window. The technique used in conducting the experiment was to fly in orbit rate, service propulsion engine forward, with the hatch window parallel to the lunar surface. Preplanned times were used to start and stop the camera, which was actuated by the 20-second intervalometer. The first pass for this experiment was accomplished with the same camera setting, but in two parts. The first part was completed for that area from approximately 10 degrees to 60 degrees sun angle, and the second part was from 60 degrees to 10 degrees. The second pass was conducted in a manner similar to the first pass, but with new camera settings and in an area near the subsolar point. No difficulties were encountered in either pass. At the completion of the multispectral photography, selected targets of opportunity, including Descartes, Fra Mauro, and the north wall of Theophilus were photographed with the same camera equipment. Digital autopilot maneuvers were conducted using ground-supplied gimbal angles, and two photographs of each area were taken. Selected targets of opportunity were photographed no closer together than approximately 5 minutes, an interval recommended as convenient for future flights, particularly where camera changes are required.

### 9.12.4 Bootstrap Photography

An additional day in lunar orbit had been planned following ascent stage deorbit to permit completion of bootstrap photography, which is so named because stereo-strip and high-resolution coverage of surface areas planned for future landings was involved. The stereostrip photography was conducted with the spacecraft longitudinal axis pointed down the lunar radius vector (local vertical) using orbit-rate torquing from the guidance system. The sextant was used for through-the-optics photography with the shaft angle set to zero and the trunnion angle to 45 degrees. In addition, the 70-mm camera, with the 80-mm lens and black-and-white film, was mounted in the right-hand rendezvous window. The strip photography was conducted using procedures outlined in the flight plan.

At the completion of the rest period at 102-1/2 hours, target-of-opportunity photographs were first taken of Fra Mauro out the right-hand window. These pictures were planned to support Apollo 13 and were taken with black-and-white film and the 80-mm lens.

High-resolution photography was obtained by using the 500-mm long-range lens and the 70-mm camera mounted on a special bracket in the right-hand rendezvous window. The crew optical sight was used for aligning the 500-mm lens. Ground-supplied gimbal angles and camera operating times were again used for this photography and subsequent landmark tracking. The high resolution photography was conducted on the areas near the craters Descartes, Fra Mauro, and Lalande, and as an additional bonus the Herschel crater area also was photographed.

Two revolutions of landmark tracking were conducted following the bootstrap photography. The telescope was used to track the target while the camera, mounted on the sextant, was used for photographic purposes. On each revolution four specified landmarks associated with future sites were tracked without difficulty.

## 9.13 TRANSEARTH INJECTION

Following a day of photography and landmark sightings, described earlier, preparation was begun for transearth injection to be conducted at the end of the 45th lunar orbit revolution. This maneuver was performed nominally using the service propulsion system. The firing duration was 2 minutes 11 seconds and residuals were trimmed to within 0.2 ft/ sec.

## 9.14 TRANSEARTH FLIGHT

Transearth coast was a fairly relaxed period for the crew. Six sets of navigation sightings were accomplished, and the techniques were the same as those used during translunar coast. A variety of stars were used, including some that were not from the standard Apollo star catalogue, to determine the effect of sighting stars and the earth when the sun is in close proximity to the earth's limb.

One exception to the attitude-control procedures was followed for the first two sets of sightings. Unbalanced couples were used in one configuration of the autopilot; that is, two adjacent reaction control quads were disabled. This procedure enabled minimum impulse with only a single thruster. The two-jet minimum impulse mode overcontrolled and would not stabilize the spacecraft, and the landmark line of sight was constantly moving. Constant minimum impulse thrusting was therefore required to keep the substellar point within the field of view. By using unbalanced couples, spacecraft motion could be nulled completely.

During transearth coast, two midcourse corrections were required. The first midcourse correction was 2 ft/sec and the second was 2.4 ft/ sec. No discrepancies were noted during either maneuver.

Soon after undocking in lunar orbit, the reacquisition mode of the high-gain antenna exhibited an anomalous behavior. This discrepancy posed no real problem because ample time was available to perform manual acquisition when necessary. During transearth coast, two tests were performed in an attempt to isolate the failure source (see section 14.1.6).

The only other event of significance during transearth coast was the observation and photography of a solar eclipse that occurred when the earth came between the spacecraft and the sun. This event was so spectacular that many photographs were taken. Because preflight planning had not accounted for this event, the crew was in doubt about the correct exposure times and camera settings.

## 9.15 ENTRY AND LANDING

Entry was normal and was conducted in accordance with the onboard checklist. The only noticeable discrepancy during entry was that, although the planned drogue deployment time was given as 8 minutes 4 seconds after entry, the actual deployment did not occur until 8 minutes 24 seconds.

Sea-state conditions were fairly rough, and the landing impact was extremely hard. (Editors note: Later information indicates the command module did not enter the water at the nominal 27.5-degree angle, from which it hangs on the parachute system. Engineering judgement indicates that the command module entered the water at an angle of 20 to 22 degrees, which corresponds to an impact acceleration of about 15g. This off-nominal condition is attributed to a wind-induced swing of the command module while it was on the parachutes and to the existing wave slope at contact.) The 16-mm sequence camera had been placed on its bracket in the right-hand rendezvous window to photograph entry but came loose at impact and contacted the Lunar Module Pilot above the right eye. Later inspection of the spacecraft revealed that portions of the heat shield had been knocked loose during impact. The spacecraft was pulled over by the parachutes to a stable II attitude. Uprighting procedures were completely adequate, and no difficulty was encountered in returning to stable I.

Recovery was nominal in all respects. Back-contamination procedures had been changed to allow the crew to wear standard blue flight suits with a portable face mask. These procedures are considered adequate and perfectly acceptable by the crew. A 10-foot static line, deployed below the retrieval net from the helicopter, actually came into the life raft and could have entangled a crewman's foot when hoisting another crewman from the raft. This hazardous line should be eliminated.

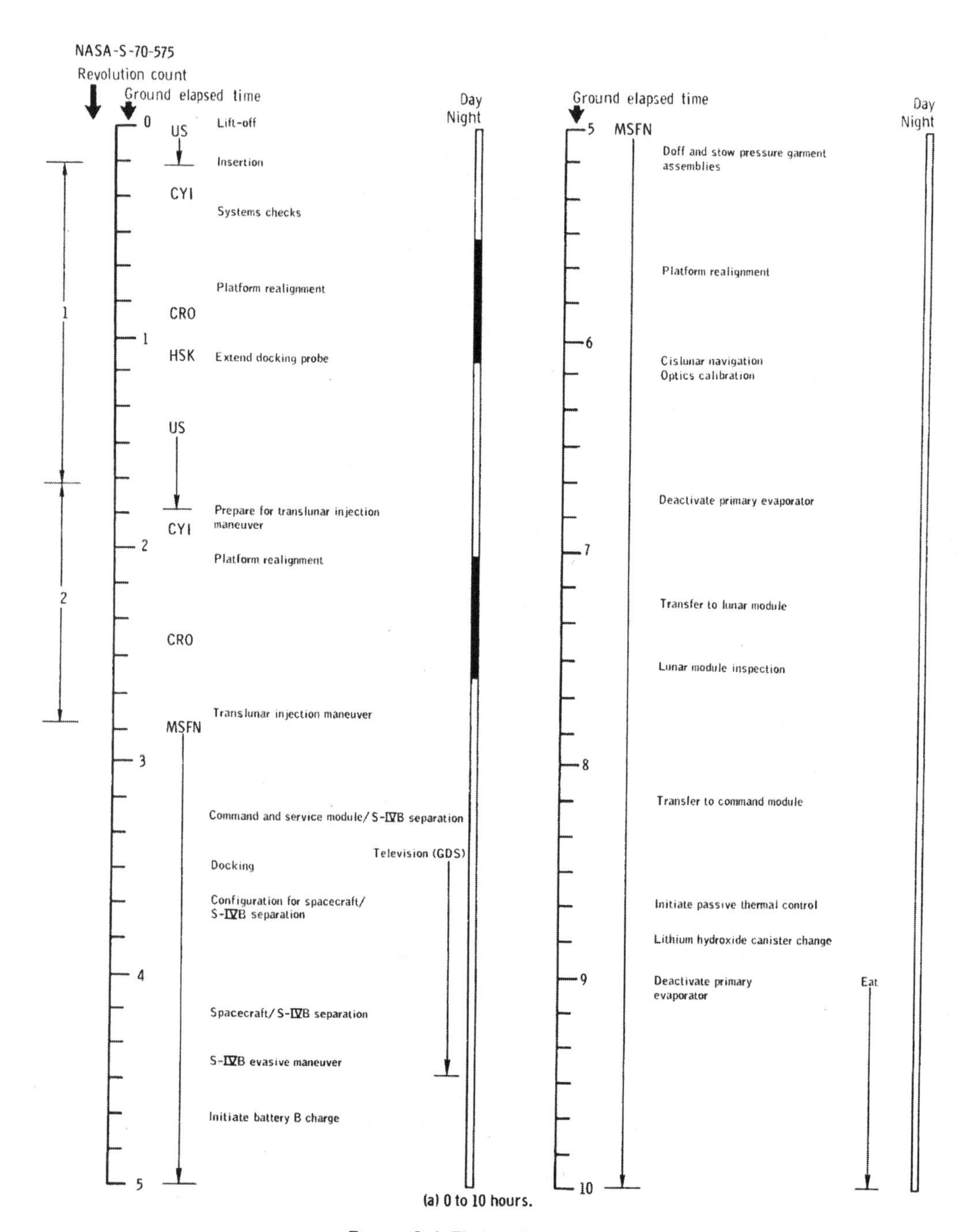

(a) 0 to 10 hours.

Figure 9-1 Flight plan activities

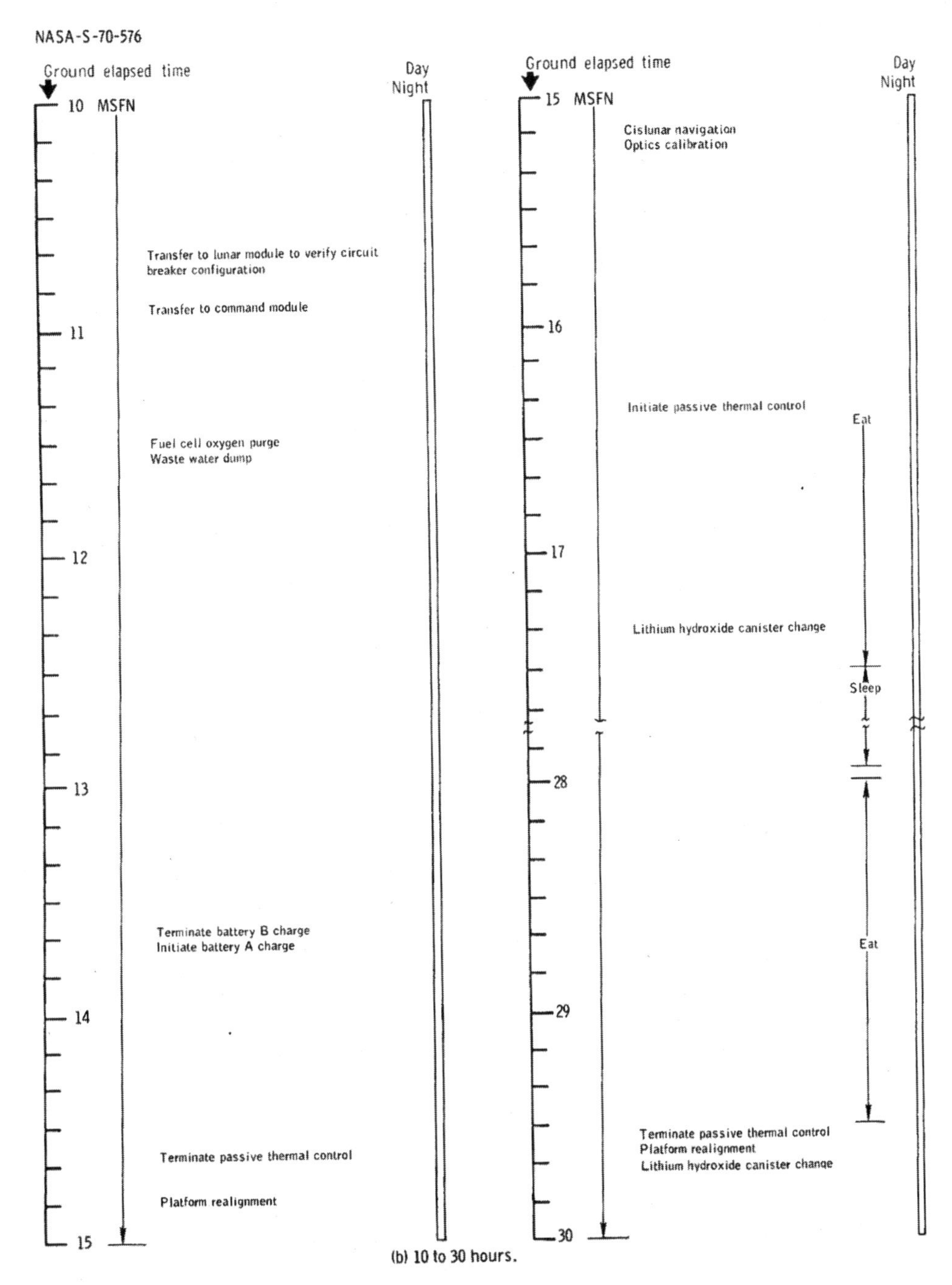


(b) 10 to 30 hours.

Figure 9-1 Continued

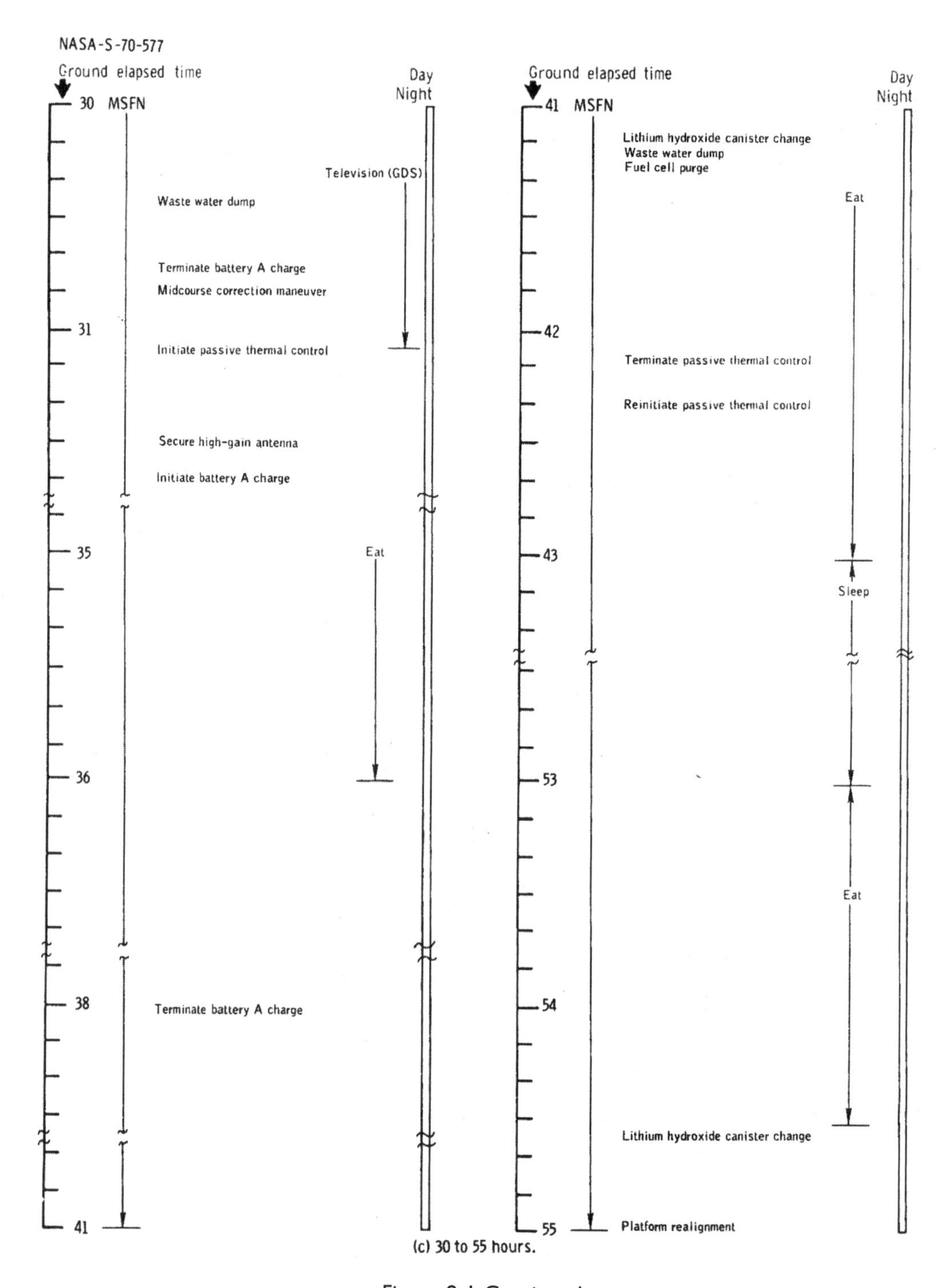

(c) 30 to 55 hours.

Figure 9-1 Continued

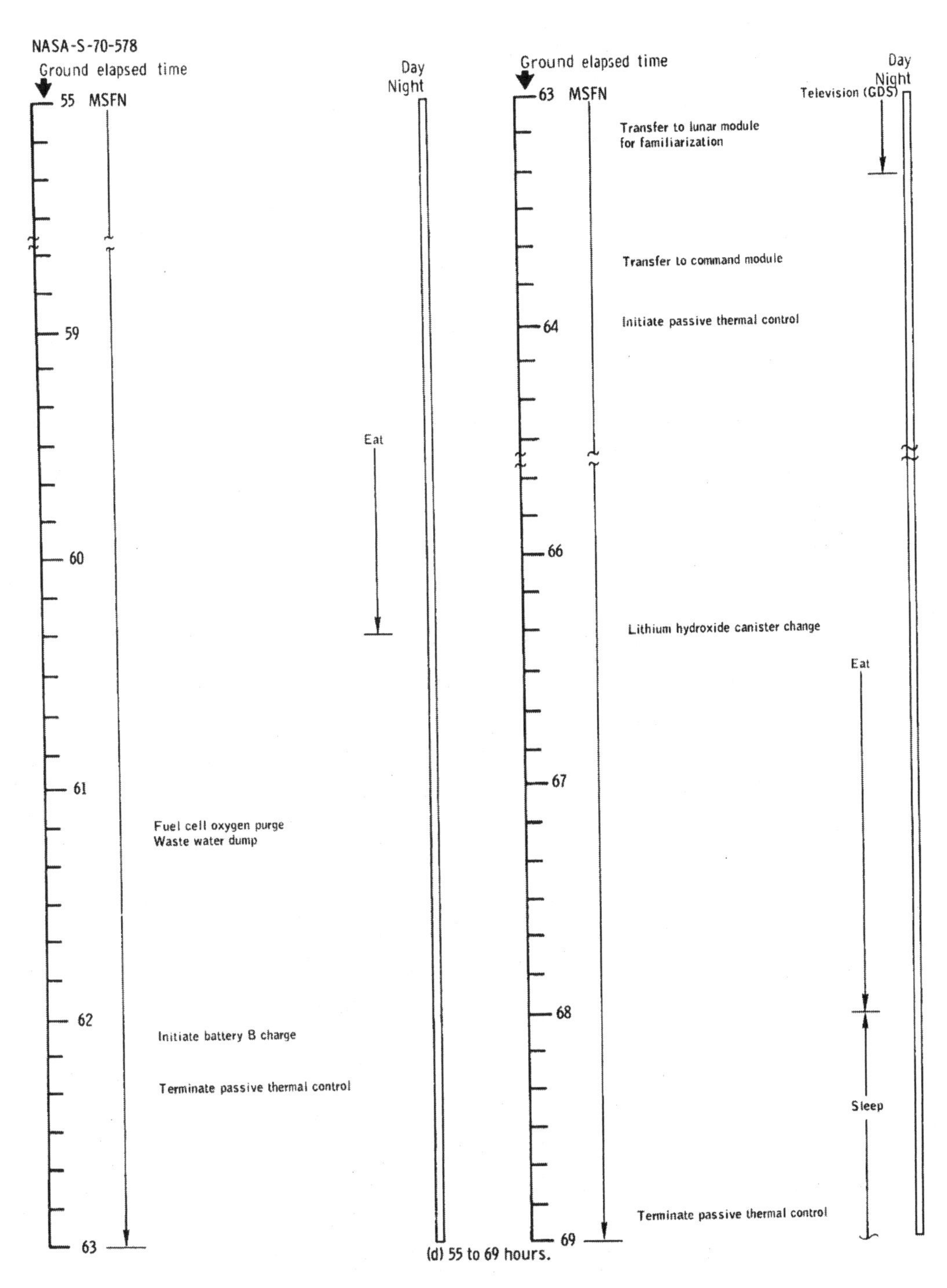


(d) 55 to 69 hours.

Figure 9-1 Continued

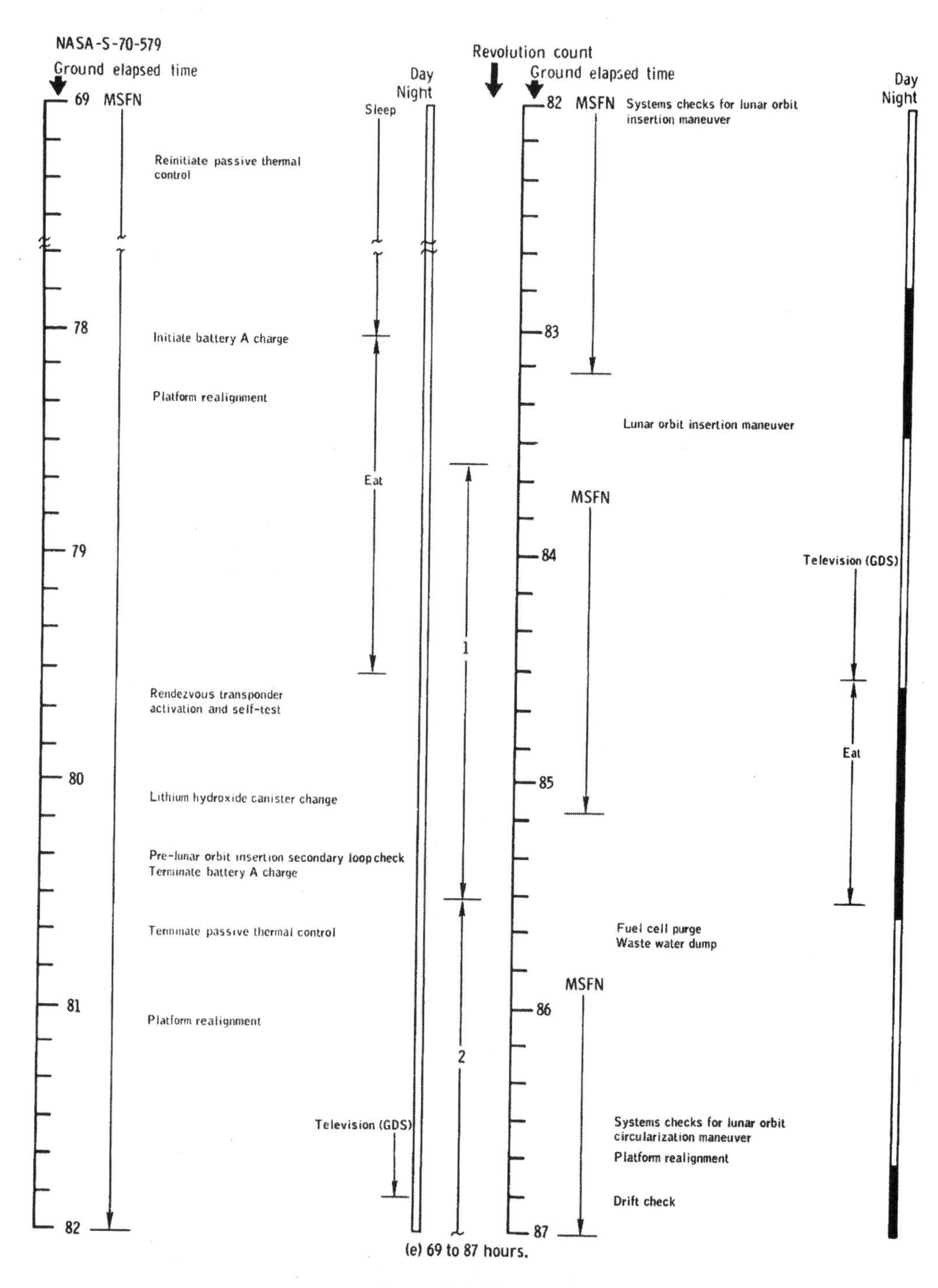


(e) 69 to 87 hours.

Figure 9-1 Continued

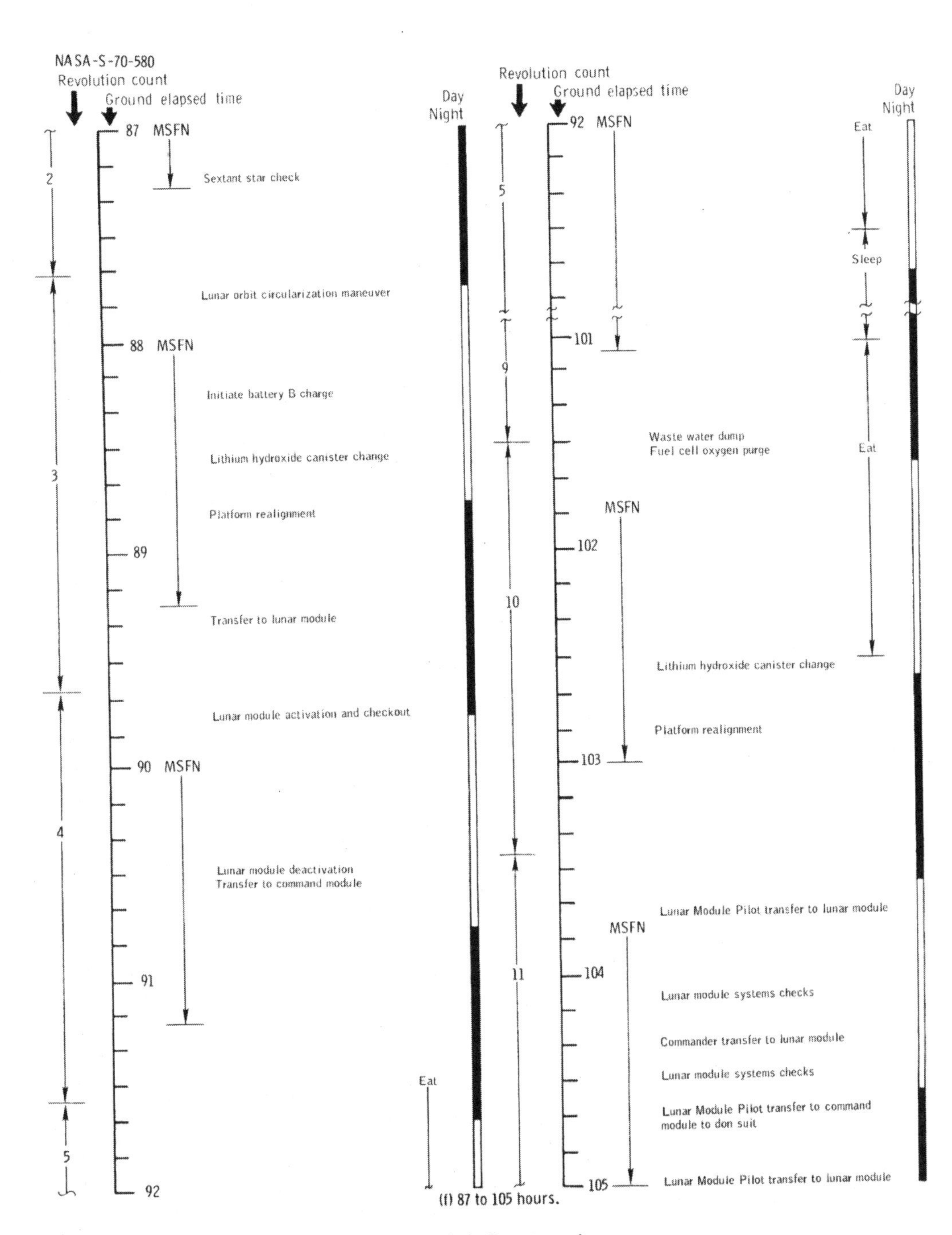


Figure 9-1 Continued

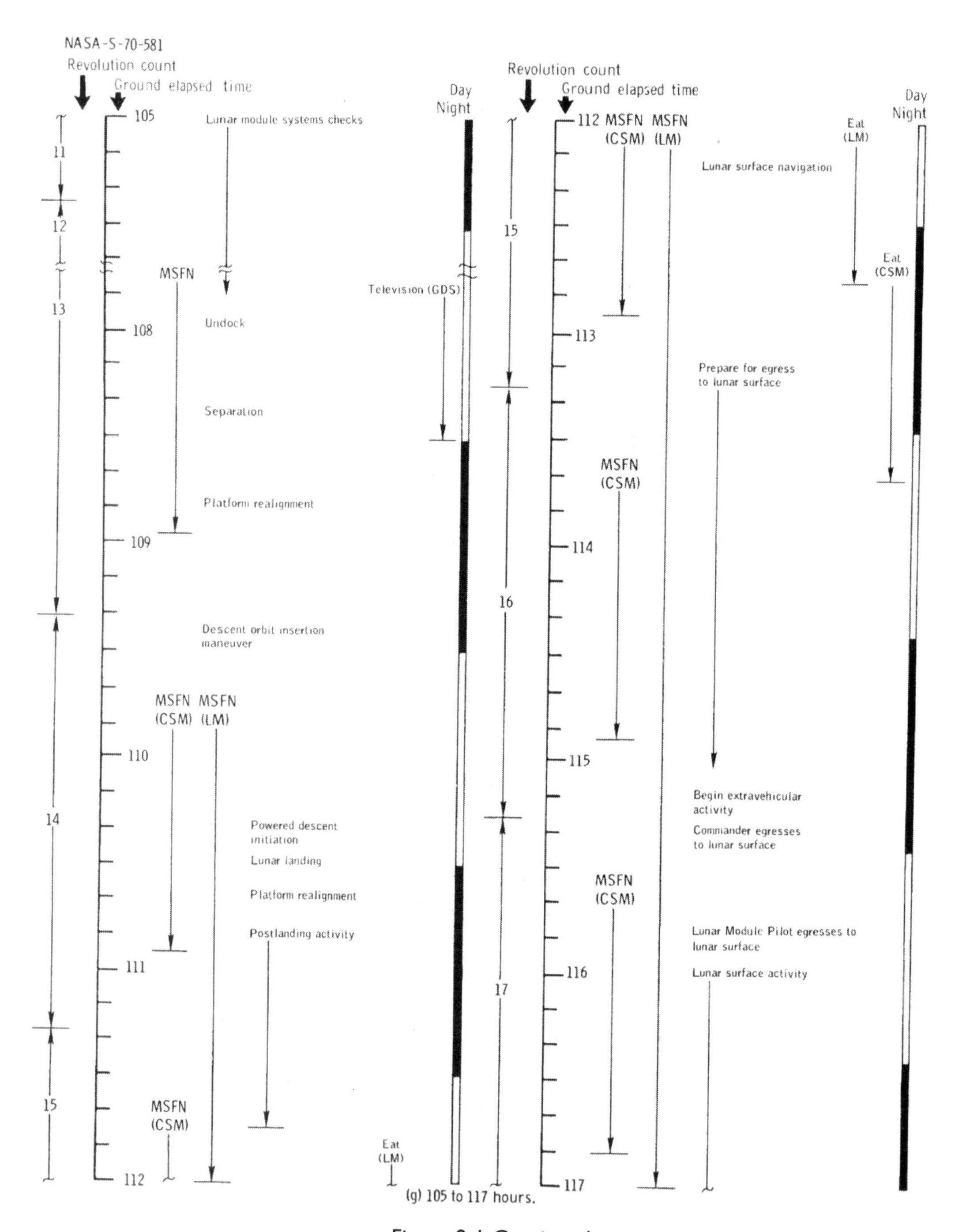


Figure 9-1 Continued

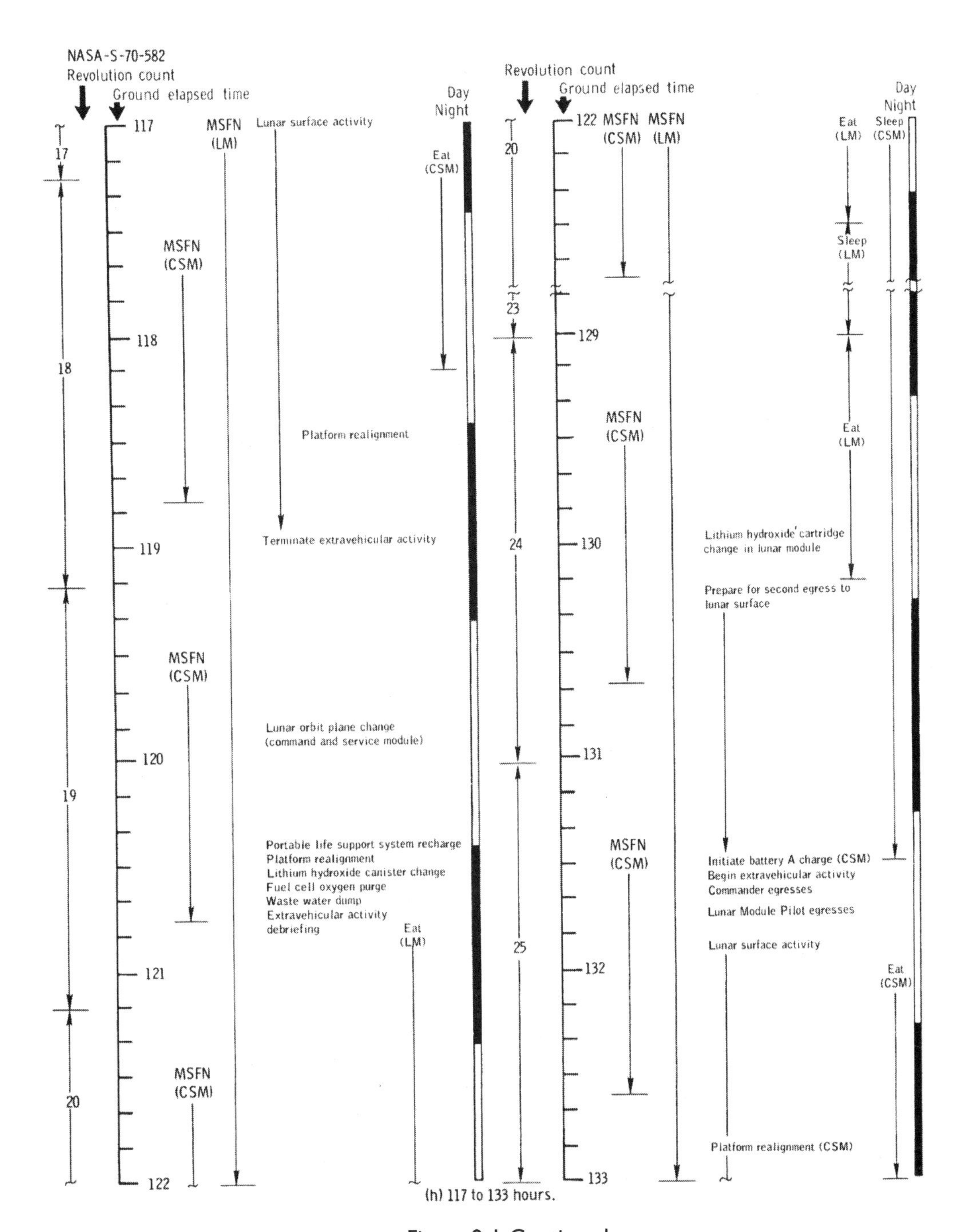


Figure 9-1 Continued

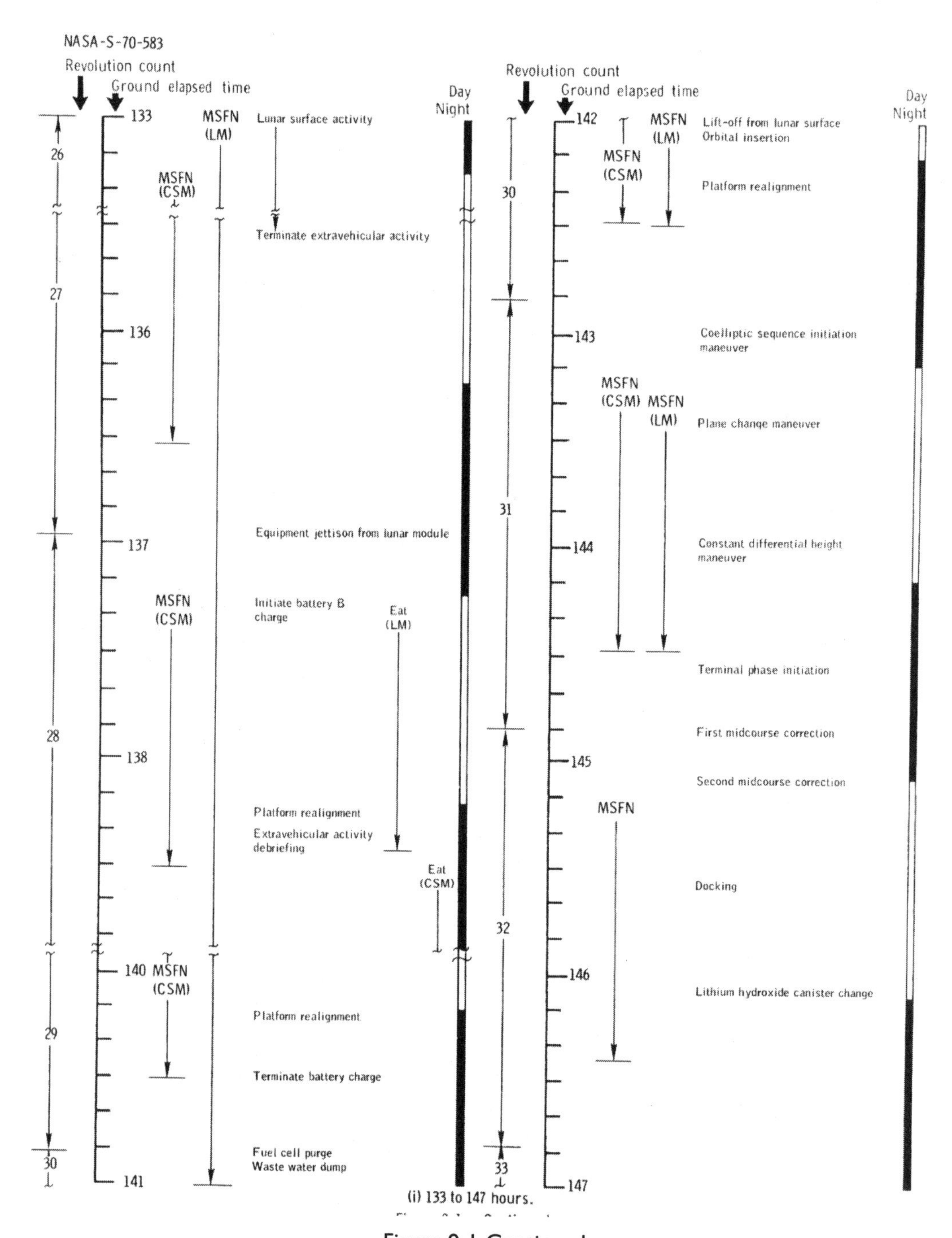


Figure 9-1 Continued

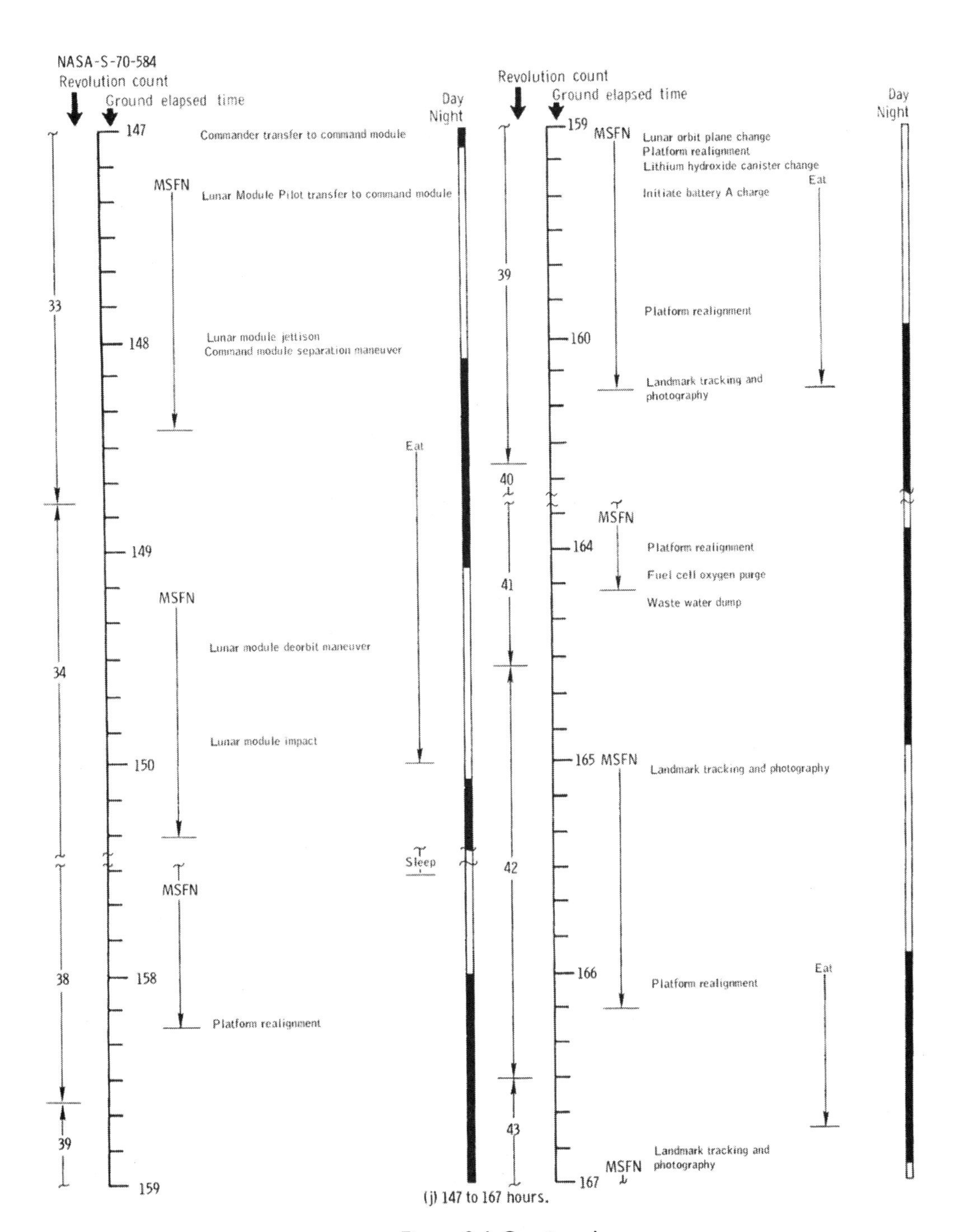

(j) 147 to 167 hours.

Figure 9-1 Continued

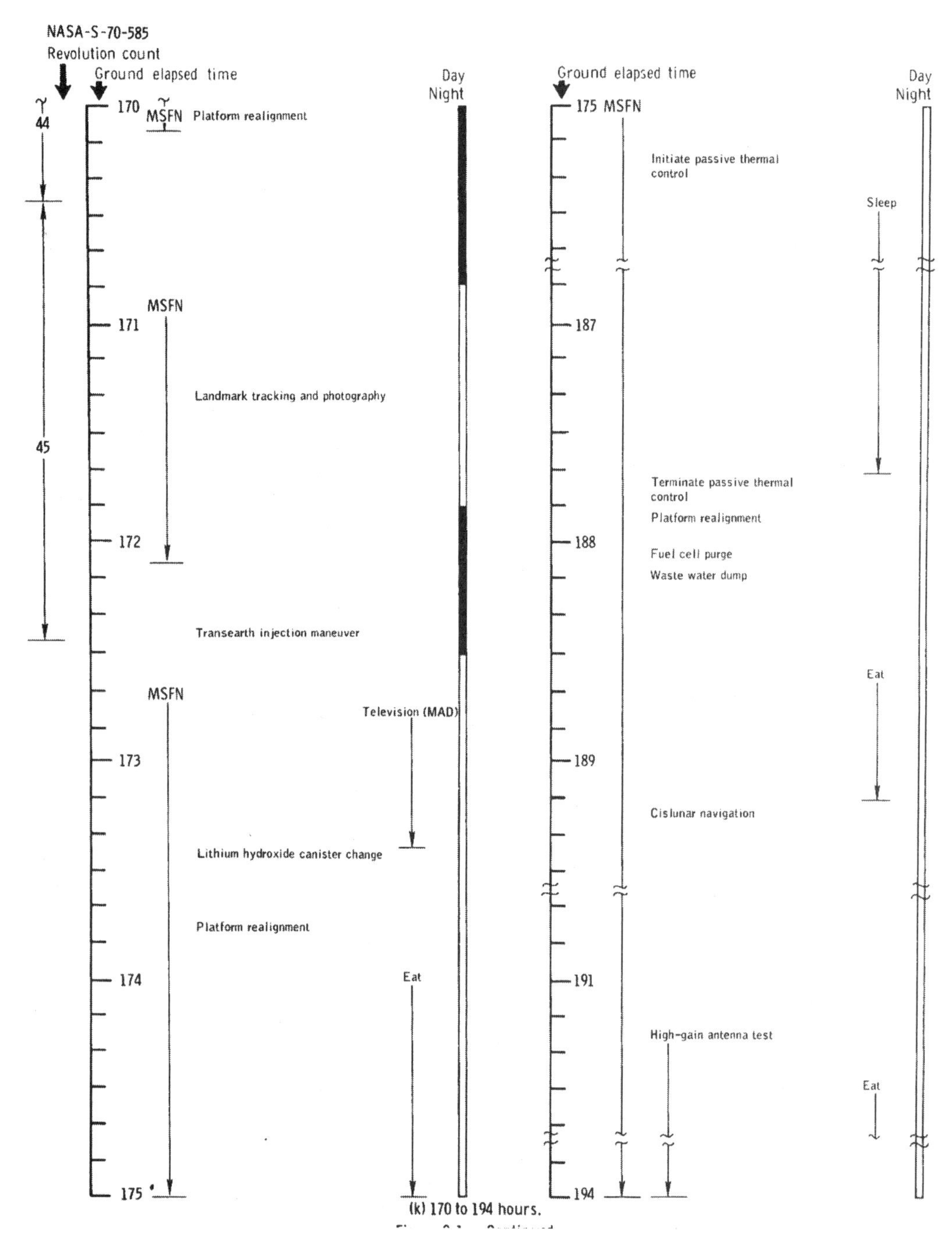


(k) 170 to 194 hours.

Figure 9-1 Continued

NASA-S-70-586

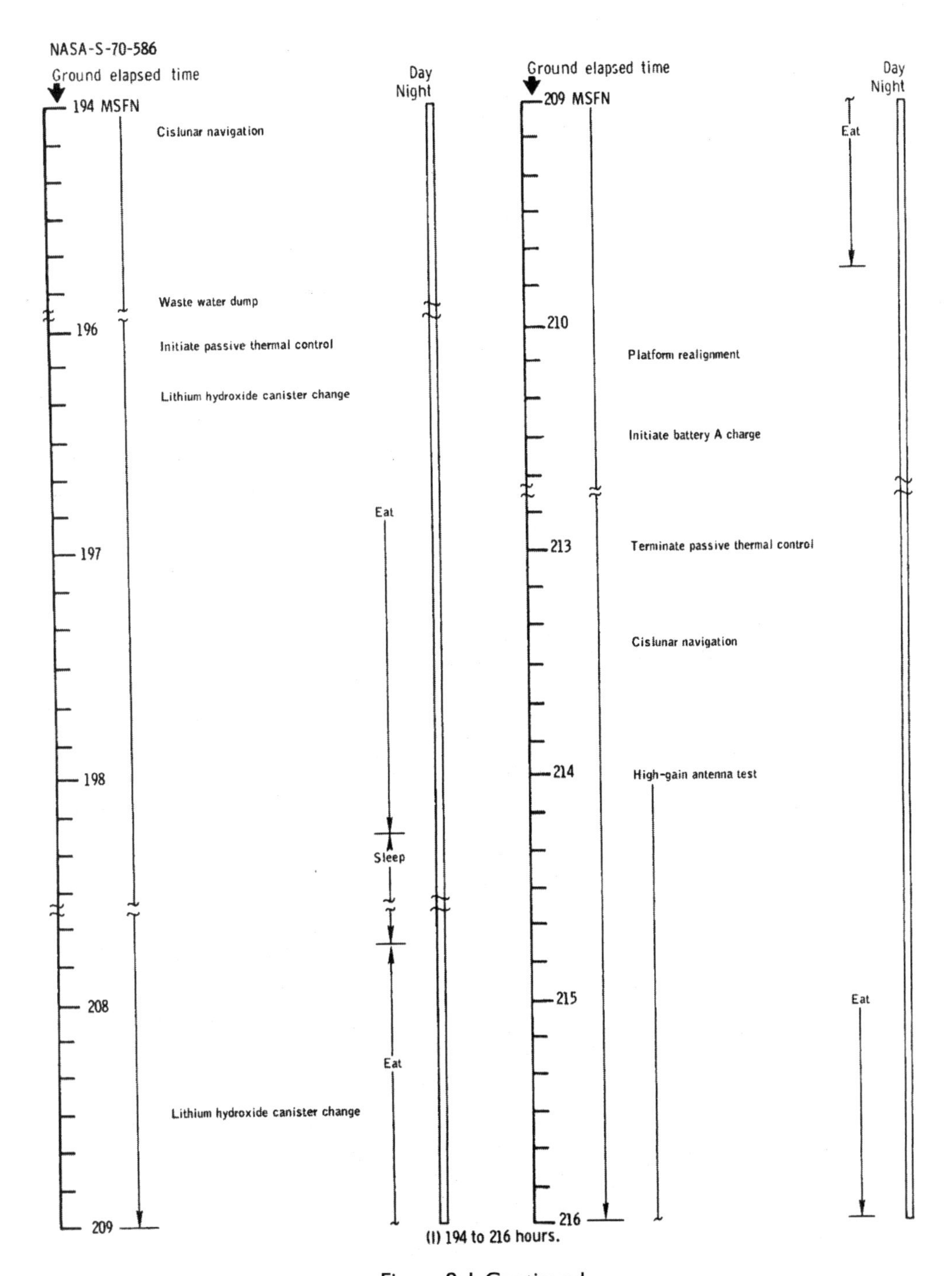


(l) 194 to 216 hours.

Figure 9-1 Continued

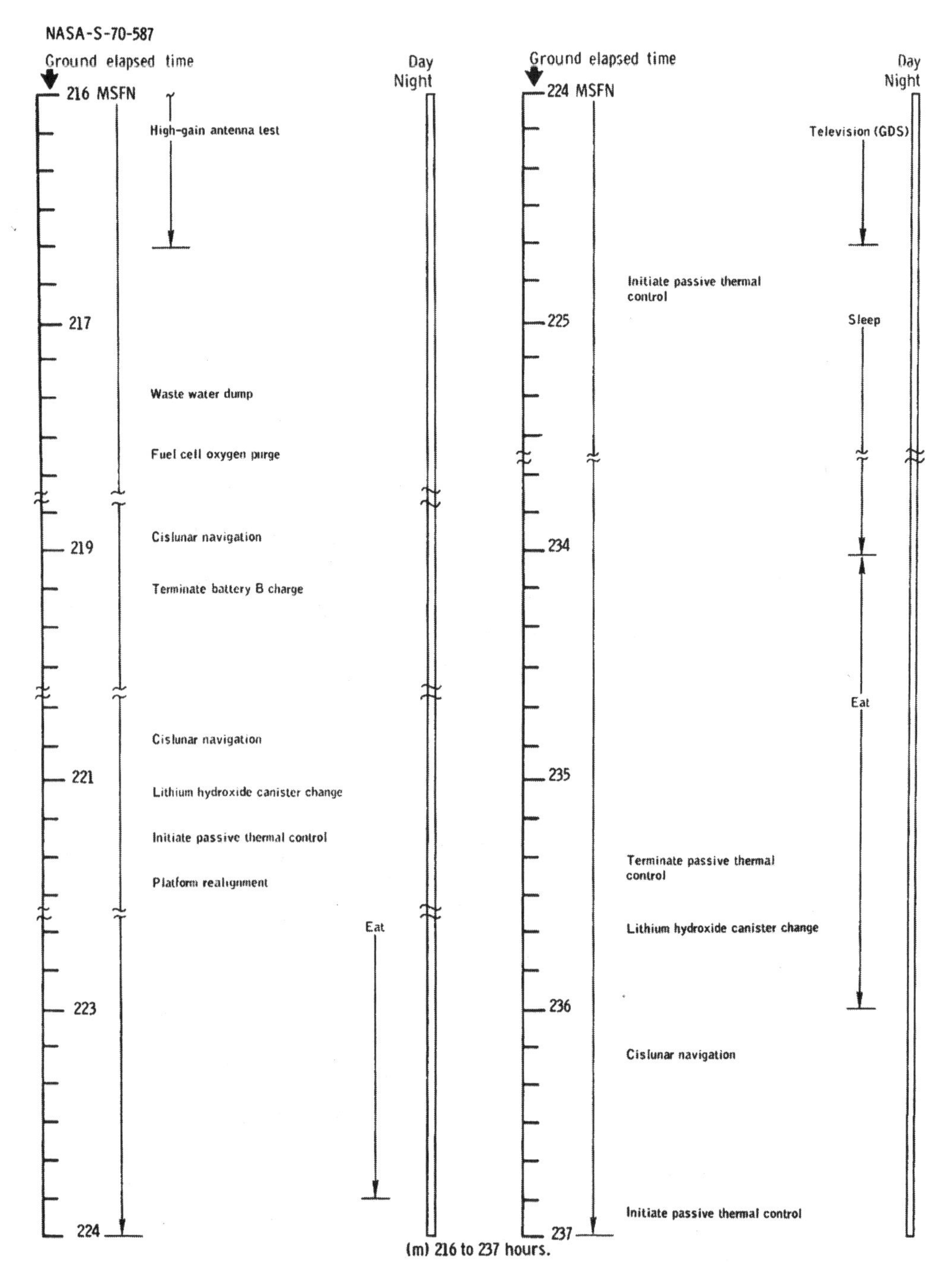


(m) 216 to 237 hours.

Figure 9-1 Continued

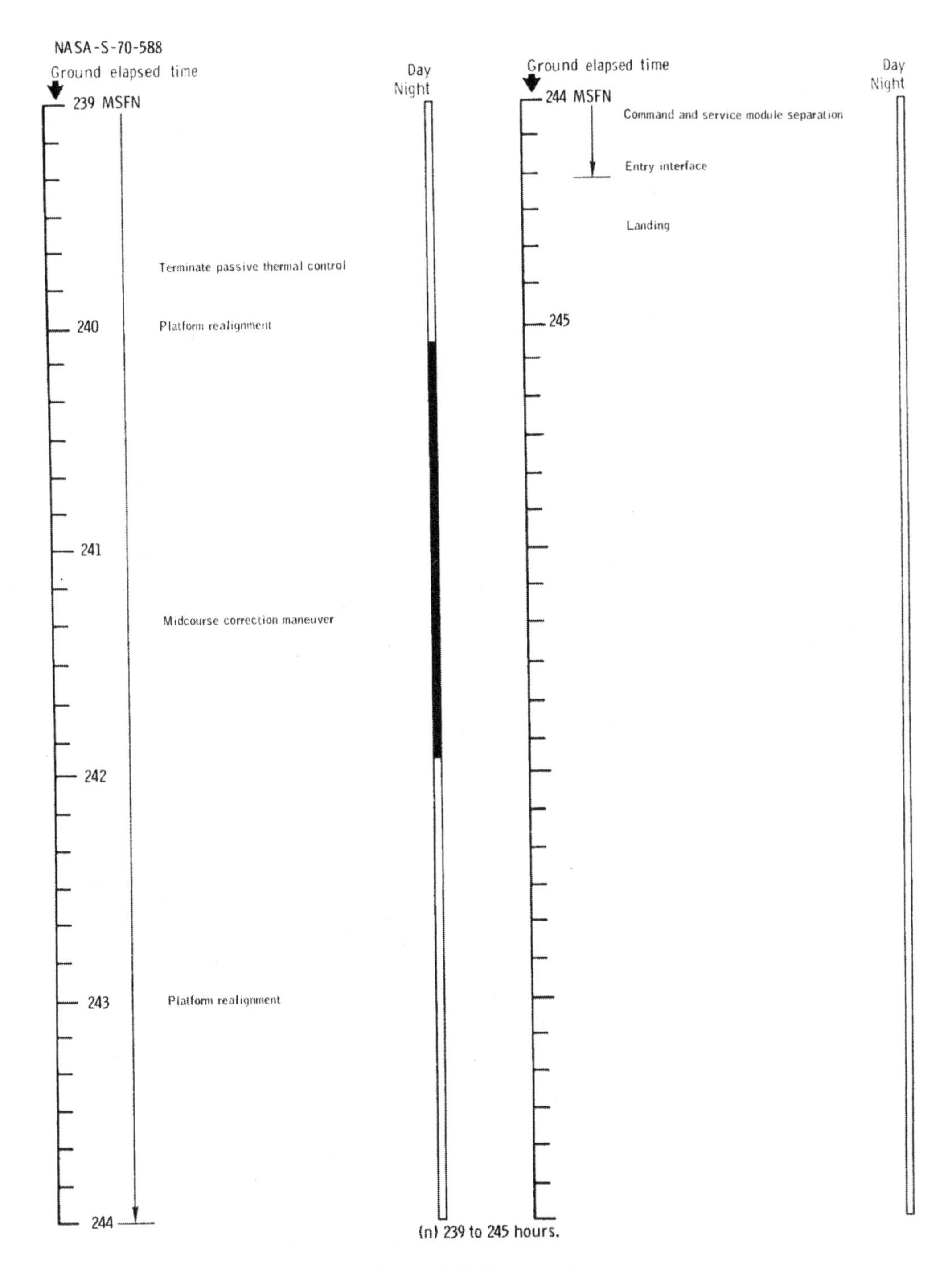


(n) 239 to 245 hours.

Figure 9-1 Continued

# 10.0 BIOMEDICAL EVALUATION

This section is a summary of Apollo 12 medical findings, based on preliminary analyses of biomedical data. More comprehensive evaluations will be published in a comprehensive medical report.

The three crewmen accumulated 734 man-hours of space flight experience during this second lunar landing mission. All inflight medical objectives were accomplished, except that sleep data on the Commander and the Lunar Module Pilot were only sporadic during the translunar coast phase.

The crew's health and performance were generally good, in spite of altered work-rest cycles. The Commander and the Lunar Module Pilot apparently became fatigued during the lunar surface stay because of inadequate rest. No adverse effects attributable to lunar surface exposure have been observed.

## 10.1 BIOINSTRUMENTATION AND PHYSIOLOGICAL DATA

Biomedical data were of good quality throughout the mission. Less than 250 hours of data were received during this 10.2-day mission, compared to 319 hours of data received during the 8.4-day Apollo 11 remission. This decrease was caused by the loss of all data from the Commander after the sixth day of the mission and by the lack of data during most sleep periods, when the crewmen elected to disconnect the biomedical umbilicals.

On the fourth day of the flight, the Commander reported that the skin under his biomedical sensors was irritated. He removed and reapplied the top sternal electrocardiogram sensor near the original application site. Upon medical recommendation, the Commander subsequently removed all sensors on the sixth day of the mission and treated the irritated skin areas with first-aid cream from the medical kit.

Just prior to lunar descent, the electrocardiogram signal from the Lunar Module Pilot became markedly degraded because the electrode paste had dried. Following the application of new electrode paste and tape, the signal was restored.

Physiological measurements were within expected ranges throughout the mission. The average heart rates for the mission were 74, 76, and 67 beats/min for the Commander, the Command Module Pilot, and the Lunar Module Pilot, respectively.

Heart rates during the two extravehicular activity periods are plotted in figures 10-1 and 10-2. The Commander's average heart rates were 74 and 108 beats/min for the first and second period, respectively; and the Lunar Module Pilot's average heart rates were 107 and 122 beats/min. After the first 30 minutes of the second period, both crewmen had sustained heart rates above 100 beats/min. The metabolic rates of each crewman during the extravehicular activities are presented in section 10.3.

## 10.2 MEDICAL OBSERVATIONS

### 10.2.1 Adaptation to Weightlessness

All crewmen reported the sensation of fullness in the head, a condition which remained for 1 or 2 days after lift-off. Their eyes were bloodshot for the first 24 hours of flight, and their faces appeared slightly rounded or swollen throughout the flight. They also reported that their shoulders tended to assume a squared-off (or raised) position, rather than being sloped in the usual relaxed position.

As in previous Apollo missions, the inflight exerciser was used primarily for crew relaxation. The crew used the exerciser several times each day for periods ranging from 15 to 30 minutes during the translunar coast.

### 10.2.2 Visual Phenomenon

The crewmen reported seeing point flashes or streaks of light. The lights were visible with the eyes both opened and closed. The crew was more aware of these flashes after retiring when they consciously tried to observe them. The Apollo 11 crew also noted occasional streaks through the cabin (discussed in reference 9). Efforts are continuing to explain this phenomenon.

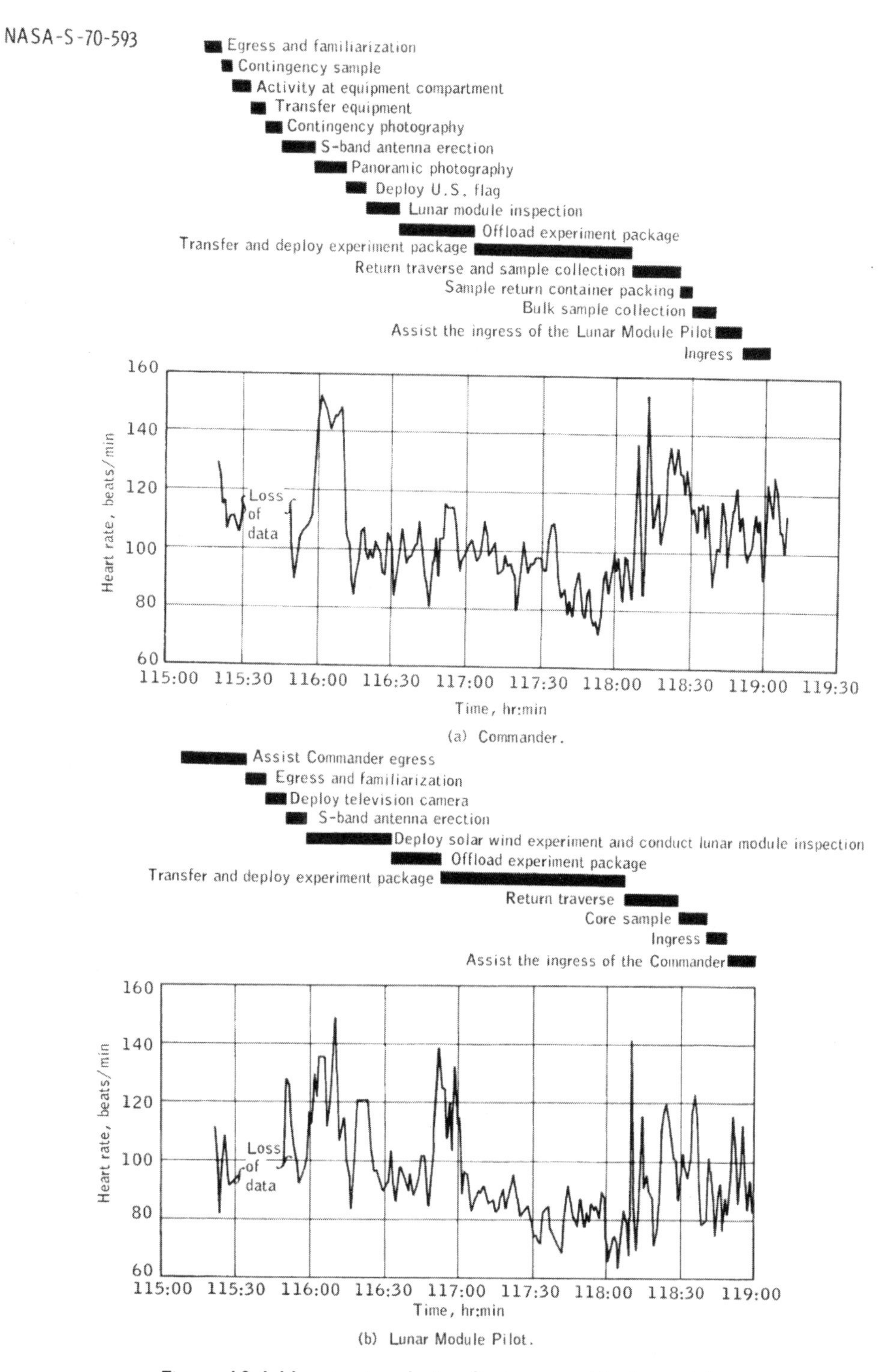


Figure 10-1 Heart rates during first extravehicular activity.

NASA-S-70-594

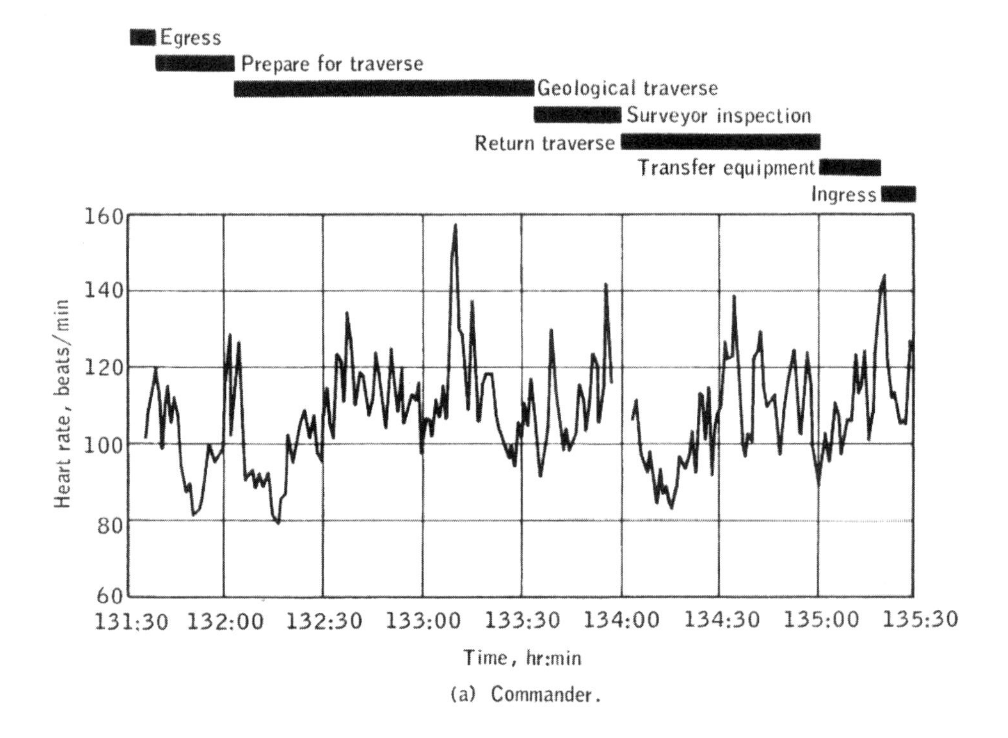


(a) Commander.

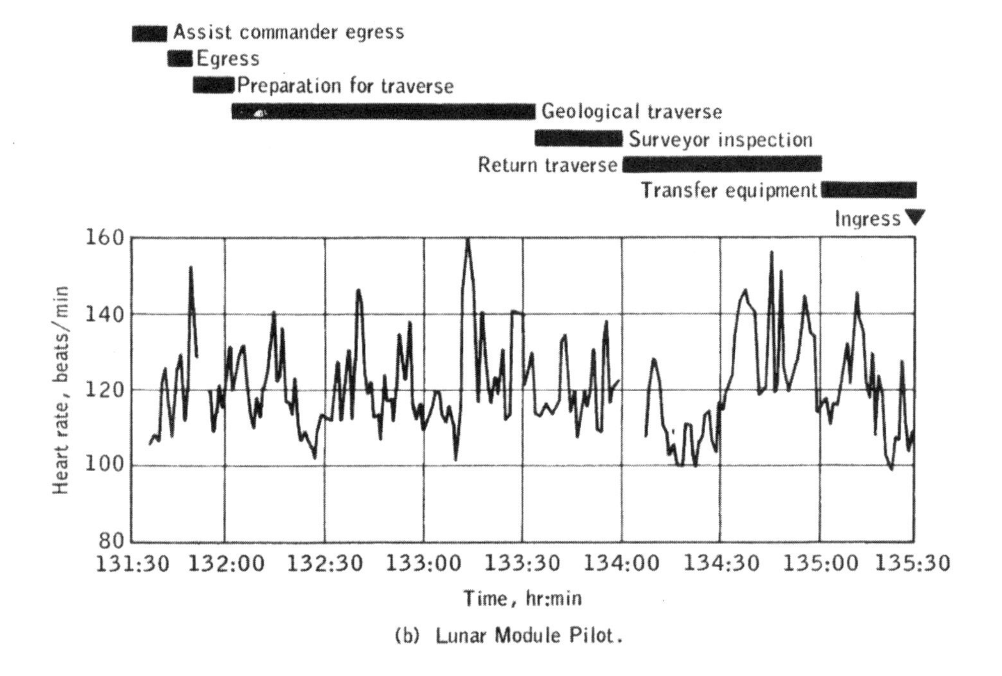


(b) Lunar Module Pilot.

Figure 10-2 Heart rates during second extravehicular activity.

### 10.2.3 Medications

All crewmen took Actifed to relieve nasal congestion at various times throughout the flight. The Lunar Module Pilot reported taking Actifed prior to lunar module descent to relieve symptoms developed after earth lift-off. The Lunar Module Pilot also took Seconal throughout most of the mission to aid sleep. Aspirin was also taken occasionally by all the crewmen. No motion sickness medications were taken prior to entry. The medication taken by each crewman follows.

| Medication | Commander | Command Module Pilot | Lunar Module Pilot |
|---|---|---|---|
| Aspirin | 2 | 2 | 2 |
| Actifed | 4 | 3 | 11 |
| Seconal | 0 | 0 | 6 |

The crewmen attempted to use the Afrin nasal spray bottles. These units were modified after Apollo 11 to contain an inner cotton pledget for preventing the rapid release of liquid when the cap was removed in zero-g. The crew said it was difficult, if not impossible, to obtain spray from these modified bottles. Postflight testing in one-g revealed that all three Afrin bottles delivered a fine spray when sharply squeezed.

Sleep periods during translunar coast began approximately 7 to 9 hours after the crew's normal bedtime of 11 p.m. The crew reported that they had no particular trouble in adapting to the shifted sleep periods. However, the first flight day was extremely long, and the crew was thoroughly fatigued by the time the first sleep period began 17 hours after lift-off.

The crewmen slept well in the command module during the translunar and transearth coast phases, and the Lunar Module Pilot took at least two unscheduled naps during transearth coast. However, they reported their sleep periods were longer than necessary, since they would invariably awaken about 1 hour ahead of time and would usually remain in their sleep stations until time for radio contact.

The lunar module crew slept only about 3 hours on the lunar surface prior to the second extravehicular activity period. In the next sleep period following rendezvous and docking, all three crewmen in the command module slept only 3 or 4 hours, which was less than desirable.

Biomedical monitoring during sleep periods was very limited. The crew complained that it was inconvenient to hook up to the biomedical harness while in the sleeping bags; hence, very little data were received.

### 10.2.5 Radiation

Initial estimates of radiation dosage were determined from the personal radiation dosimeters worn by each of the crew and from the Van Allen belt dosimeter. The final readings from the personal radiation dosimeters yielded net integrated (uncorrected) doses of 690, 630, and 640 mrad for the Commander, the Command Module Pilot, and the Lunar Module Pilot, respectively. The Van Allen belt dosimeter displayed integral doses of 510-mrad depth dose and 970-mrad skin dose for the command module. The personal radiation dosimeters and the Van Allen belt dosimeter skin-dose sensor did not integrate comparable doses during the return passage through the Van Allen belts, although it was predicted that the readings would be nearly equal. The possibility exists that the personal dosimeters were stowed in a way that increased radiation shielding.

Approximately half of the total dose recorded on the personal radiation dosimeters was received during the phase just prior to entry. This disparity was expected because of a different trajectory which resulted in a longer traverse through the Van Allen belts.

The crewmen were examined under total body gamma spectroscopy following release from quarantine on December 10, 1969. The preliminary analysis revealed no induced radioactivity.

### 10.2.6 Water

The crew reported that the drinking water in both the command module and the lunar module was most satisfactory. The nine inflight chlorinations of the command module water system were accomplished as scheduled in the flight plan. Analysis of water from the hot-water port approximately 14.5 hours after splashdown, or 35.5 hours after the last inflight chlorination, showed a free-chlorine residual of 0.125 mg/l. A postflight analysis of water from the drink gun was not performed. Preflight testing showed that the iodine level in the lunar module water tanks was adequate for bacterial protection throughout the flight.

Chemical and microbiological analyses of the preflight water samples for the command module showed no significant contaminants. The pH concentration of the lunar module water was uniformly low in preflight testing, and the nickel ion concentrations were slightly elevated in the final water load after iodination. However, the low pH and the elevated nickel ion concentrations are not considered medically significant for flights on the order of 1 or 2 weeks in duration.

### 10.2.7 Food

The food supply was very similar to that for Apollo 11. The two new foods included in the menu for this mission were rehydratable scrambled eggs and wet-pack beef and gravy. Maximum use was made of the spoon-bowl packages for the various rehydratable food items, and the spoon size was increased from one teaspoon to one tablespoon. The pantry-type food system, which allows open selection of all food items was again used for this mission. Four meal periods on the lunar surface were scheduled, and extra optional items were included with the normal meal packages.

Prior to the flight, each crewman evaluated the available food items and selected his individual menu. These menus provided approximately 2300 kilocalories per man per day. The crew made an effort to follow the menus and to maintain the onboard log of foods consumed. Favorable comments were received about the quality of the food throughout the flight. After the flight, the crew reported that gas in the hot-water supply tended to inhibit complete rehydration of food. Some of the gas was removed by opening the spoon-bowl packages and mixing the food with a spoon. No package failures were experienced. The crew had no difficulty eating any of the food items with a spoon.

## 10.3 EXTRAVEHICULAR ACTIVITIES

The integrated metabolic rates and the accumulated work production during the planned activities are listed in tables 10-I and 10-II. Heart rates during the extravehicular periods are plotted in figures 10-1 and 10-2. The predicted and actual metabolic productions follow.

| | Metabolic production, Btu/hr | | | |
|---|---|---|---|---|
| Crewman | First period | | Second period | |
| | Observed | Predicted | Observed | Predicted |
| Commander | 975 | 1166 | 875 | 1210 |
| Lunar Module Pilot | 1000 | 1142 | 1000 | 1134 |

## 10.4 PHYSICAL EXAMINATIONS

Comprehensive physical examinations were conducted on each crewman at 30, 14, and 5 days prior to launch. Brief examinations were conducted daily on the last 5 days before launch, and a comprehensive examination was conducted immediately after recovery.

The recovery day physical examinations revealed that the crewmen were in good health. Body temperatures were normal, and body weights were within expected values. The Lunar Module Pilot had a small amount of clear fluid with air bubbles in the middle ear cavity, but this symptom disappeared after 24 hours of decongestant therapy. Because the command module splashed down normal to the surface of the water, landing forces were greater than those experienced on previous Apollo flights. A camera came off the window bracket and struck the Lunar Module Pilot on the forehead. He lost consciousness for about 5 seconds and sustained a 2-centimeter laceration over the right eyebrow. The cut was sutured soon after retrieval and healed normally.

TABLE 10-I.- METABOLIC ASSESSMENT OF FIRST EXTRAVEHICULAR ACTIVITY

| Surface activity | Starting time, Min | Duration, hr:min | Metabolic rate, Btu /hr | Estimated work, Btu | Cumulative work, Btu |
|---|---|---|---|---|---|
| | | Commander | | | |
| Extravehicular preparation | 115:14 | 2 | 350 | 11 | 11 |
| Egress | 115:16 | 6 | 1250 | 124 | 135 |
| Environmental familiarization | 115:22 | 3 | 1250 | 62 | 197 |
| Contingency sample collection | 115:25 | 5 | 1100 | 92 | 289 |
| Equipment bag transfer | 115:30 | 16 | 1200 | 317 | 606 |
| Contingency photography | 115:46 | 6 | 1050 | 108 | 714 |
| S-band antenna deployment | 115:52 | 18 | 1250 | 372 | 1086 |
| U.S. flag deployment | 116:10 | 10 | 950 | 162 | 1248 |
| Panoramic photography | 116:20 | 12 | 800 | 169 | 1417 |
| Unload experiment package | 116:32 | 20 | 800 | 266 | 1683 |
| Transfer experiment package | 116:52 | 9 | 1000 | 148 | 1831 |
| Deploy experiment package | 117:01 | 59 | 700 | 686 | 2517 |
| Return traverse | 118:00 | 27 | 1050 | 468 | 2985 |
| Sample container packing | 118:27 | 25 | 1250 | 526 | 3511 |
| Equipment transfers | 118:52 | 10 | 950 | 165 | 3676 |
| Ingress | 119:02 | 6 | 1300 | 128 | 3804 |
| TOTAL | | 234 | 975* | | 3804 |
| | | Lunar Module Pilot | | | |
| Safety monitoring | 115:14 | 35 | 1050 | 615 | 615 |
| Egress | 115:14 | 3 | 1225 | 61 | 676 |
| Television deployment | 115:52 | 18 | 1050 | 317 | 993 |
| Deploy solar wind experiment | 116:10 | 5 | 1000 | 92 | 1085 |
| Lunar module inspection | 116:15 | 17 | 1225 | 347 | 1432 |
| Unload experiment packawc | 116:32 | 20 | 1075 | 360 | 1792 |
| Transfer experiment package | 116:52 | 9 | 1450 | 216 | 2008 |
| Activate experiment packaE.t: | 117:01 | 59 | 775 | 777 | 2785 |
| Return traverse | 118:00 | 35 | 1050 | 616 | 3401 |
| Core-tube sample | 118:35 | 16 | 925 | 249 | 3650 |
| Ingress | 118:51 | 1 | 1275 | 20 | 3670 |
| Safety monitoring | 118:52 | 16 | 850 | 230 | 3900 |
| TOTAL | | 234 | 1000* | | 3900 |

*Average

**TABLE 10-II.- METABOLIC ASSESSMENT OF SECOND EXTRAVEHICULAR ACTIVITY**

| Surface activity | Starting time, Min | Duration, hr :min | Metabolic rate, Btu/hr | Estimated work, Btu | Cumulative work, Btu |
|---|---|---|---|---|---|
| | Commander | | | | |
| Extravehicular preparation | 131:35 | 2 | 500 | 16 | 16 |
| Egress | 131:37 | 2 | 1250 | 41 | 57 |
| Equipment bag transfer | 131:39 | 5 | 850 | 70 | 127 |
| Traverse preparations | 131:44 | 16 | 650 | 173 | 300 |
| Initial geological traverse | 132:00 | 83 | 875 | 1220 | 1520 |
| Core-tube sampling | 133:23 | 13 | 850 | 185 | 1705 |
| Final geological traverse | 133:36 | 17 | 900 | 255 | 1960 |
| Surveyor inspection | 133:53 | 41 | 825 | 570 | 2530 |
| Return to spacecraft | 134:34 | 12 | 1050 | 211 | 2741 |
| Sample container packing | 134:46 | 25 | 900 | 377 | 3118 |
| Equipment transfers | 135:11 | 9 | 875 | 131 | 3249 |
| Ingress | 135:20 | 3 | 1500 | 74 | 3321 |
| TOTAL | | 228 | 875* | | 3321 |
| | Lunar Module Pilot | | | | |
| Safety monitoring | 131:35 | 9 | 875 | 131 | 131 |
| Egress | 131:44 | 5 | 1150 | 95 | 226 |
| Contrast chart photography | 131:49 | 22 | 975 | 356 | 582 |
| Initial geological traverse | 132:11 | 72 | 975 | 1166 | 1748 |
| Core-tube sampling | 133:23 | 13 | 1075 | 232 | 1970 |
| Final geological traverse | 133:36 | 17 | 975 | 274 | 2244 |
| Surveyor inspection | 133:53 | 41 | 950 | 645 | 2889 |
| Return to spacecraft | 134:34 | 12 | 1275 | 254 | 3143 |
| Closeup photography | 134:46 | 22 | 1100 | 402 | 3545 |
| Ingress | 135:08 | 3 | 1300 | 66 | 3611 |
| Equipment transfers | 135:11 | 12 | 925 | 183 | 3794 |
| TOTAL | | 228 | 1000* | | 3794 |

*Average

All crewmen suffered varying degrees of skin irritation at the biomedical sensor sites. The Command Module Pilot's skin condition was the worst of the three on recovery day. He had multiple pustules at the margins and in the center of the sensor sites. Healing lesions were noted on the Commander's skin at all sensor sites. He had removed his sensors 4 days prior to recovery and had cleansed the skin and applied cream to the affected areas daily. Red areas and small pustules were noted about all sensor sites on the Lunar Module Pilot.

The skin reaction to the sensors was the most severe seen in manned flight; therefore, a study was initiated to determine the cause of the skin irritation. The results disclosed that the Commander was allergic to some, as yet unidentified, substance in the flight electrode paste, while the other two crewmen developed no allergic reaction during these tests. Chemical analysis of the paste was inconclusive in determining the cause of the irritation. No bacteria were cultured from the electrode paste, which contains a substance to inhibit the growth of bacteria. There was a heavy concentration of Staphylococcus aureus, cultured from the skin of all three crewmen after the flight. This bacteria could account for the inflammation of the irritated skin area reported.

On the day after recovery, the Commander developed a left maxillary sinusitis which was treated successfully with decongestants and antibiotics.

Examinations were conducted daily in the Lunar Receiving Laboratory during the quarantine period, and the immuno-hematology and microbiology revealed no changes attributable to lunar material exposure.

### 10.5 LUNAR CONTAMINATION AND QUARANTINE

The procedures for quarantine of the crew and the equipment exposed to lunar material and the measures for the prevention of back contamination are discussed in reference 9. The medical aspects of lunar dust contamination are briefly discussed in section 6.

#### 10.5.1 Recovery Procedures

During recovery and return of the crew and the command module to the Lunar Receiving Laboratory, no violations of the quarantine procedures occurred. These procedures were essentially the same as for Apollo 11, with the following exceptions.

a. The biological isolation garments were not used, since they proved to be uncomfortably hot during recovery operations. They were replaced with lightweight coveralls and biological masks, which filtered the exhaled air.

b. The tunnel from the mobile quarantine facility to the command module used an improved pressure seal in the area around the hatch. Tape, which provided a successful seal when intact but could be easily pulled off, had been used to seal off the command module for Apollo 11. The pressure seal for Apollo 12 satisfactorily isolated the command module interior and no leaks occurred.

#### 10.5.2 Quarantine

A total of 28 persons, including the crew and members of the medical support teams, were exposed, directly or indirectly, to the lunar material and were subsequently quarantined in the Lunar Receiving Laboratory. Daily medical observations and periodic laboratory examinations showed no signs of infectious disease related to lunar exposure. No significant trends were noted in any biochemical, immunological, or hematological parameters in either the flight crew or the medical support personnel. The personnel quarantined in the crew reception area of the Lunar Receiving Laboratory were approved for release from quarantine on December 10, 1969. The spacecraft and samples of lunar material stored in the Lunar Receiving Laboratory were released soon thereafter.

## 11.0 MISSION SUPPORT PERFORMANCE

### 11.1 FLIGHT CONTROL

Flight control performance was satisfactory in providing operational support. Some spacecraft problems were encountered and evaluated, most of which are discussed elsewhere in this report. Only those problems which particularly influenced flight control operations or resulted in significant changes to the flight plan are discussed.

As a result of the lightning incidents which caused a power switchover and loss of platform reference during launch, several additional systems checks were conducted during earth orbit to verify systems operation prior to translunar injection. Also, an early checkout of lunar module systems was made after ejection. Lunar module power remained on for approximately 24 minutes, and no problems were discovered during this inspection. The earth orbit operations recommended specifically because of the power switchover and platform loss were as follows:

a. At insertion, the two inertial platform circuit breakers were pulled to remove power from the platform gyros and allow the gyros to spin down, terminating the tumbling of the platform gyros. The breakers were reset after 3 minutes, and the platform was aligned using an appropriate computer program during the first night pass. A new reference matrix was uplinked to the computer from the Canary Islands station, which had to be reconfigured from S-IVB to command module support. A platform realignment was performed during the second night pass to check gyro drift and verify that the lightning which caused the platform loss had not resulted in permanent damage.

b. An erasable memory dump was performed over the Carnarvon station to verify that the potential discharges had not altered the computer memory.

c. A new state vector was uplinked because the spacecraft had lost its state vector when platform reference was lost.

d. A computer self-test, a thrust vector control check, and a gimbal drive check were performed to verify spacecraft operation for a safe abort to earth, if required.

e. A new battery charging plan was transmitted to compensate for the battery power usage while the fuel cells were off the line during launch.

Following completion of the lunar module inspection and return to the command module, the lunar module current was found to be 1 ampere higher than expected. The floodlight switch on the lunar module hatch was believed to have malfunctioned, causing the floodlights to remain on. An entry into the lunar module was then required to pull the floodlight circuit breaker, and no further problems were encountered (section 14.2.1). See section 14.1.3 for a complete discussion of the launch phase discharge anomaly.

Voice interference on the lunar module downlink appeared during the first extravehicular activity. An investigation was conducted of active network sites to assure there was no network problem. The problem did not recur after this extravehicular period except for 12 seconds during the second extravehicular activity period.

## 11.2 NETWORK PERFORMANCE

The Mission Control Center and the Manned Space Flight Network provided excellent support throughout the mission. Only minor problems were encountered with computer hardware at the Mission Control Center and communication processors at the Goddard Space Flight Center.

The Carnarvon station experienced a computer hardware failure and was required to support translunar injection without command capability. During transearth coast, data were lost for 8 minutes when the spacecraft antennas could not be switched because of a command computer problem at Goldstone. After the first extravehicular activity period, a 2-kHz tone was present in the received air-to-ground communications in the lunar module backup voice mode. This tone was being generated in equipment at the Madrid station, uplinked to the lunar module, and retransmitted to the ground transponder.

## 11.3 RECOVERY OPERATIONS

The Department of Defense provided the recovery support commensurate with the probability of landing within a specified area and with any special problems associated with such a landing. The recovery force deployment is detailed in table 11-I.

Support for the primary landing area in the Pacific Ocean was provided by the antisubmarine aircraft carrier USS Hornet and eight aircraft. One of the E-1B aircraft was designated as "Air Boss," and the second as a communications relay aircraft. A third E-1B aircraft was serving as a backup and could have assumed either the "Air Boss" or a communications relay function. Two of the SH-SD helicopters, designated as "Swim 1" and "Swim 2," carried swimmers and the required recovery equipment. The third helicopter was used as a photographic platform and the fourth, designated "Recovery," carried the decontamination swimmer and the flight surgeon and was utilized for crew retrieval. A fifth helicopter was available as a backup.

**TABLE 11-1- RECOVERY SUPPORT**

| Landing area | Maximum retrieval time, hr | Maximum access time, hr | Support Number | Unit | Remarks |
|---|---|---|---|---|---|
| Launch site | | | 1 | LCU | Landing craft utility (landing craft with command module retrieval capability) |
| | | | 1 | HH-3E | Helicopter with para-rescue team |
| | | 1/2 | 2 | HH-53C | Helicopters capable of lifting the command module; each with para-rescue team |
| | | | 1 | ATF | USS Salinan |
| | | | 2 | SH-3D | Helicopters with SOA-13 Sonar |
| Launch abort | 24 in Sector A, no maximum in Sector B | 4 | 1 | DD | USS Hawkins |
| | | | 3 | HC-130 | Fixed wing aircraft; one each staged from Kindley AFB, Bermuda; from Pease AFB, N. M.; and from Lajes AFB, Azores |
| Earth orbit secondary | 24 | 6 | 2 | DD | USS Hawkins and USS Strauss |
| | | | 4 | HC-130 | Two each at Kindley AFB and at Hickam AFB, Hawaii |
| Deep space secondary | 24 | | 1 | LPH | USS Austin |
| | | | 1 | CVS | USS Hornet |
| | | 14 | 4 | SH-3D | Helicopters, 2 with swimmers, 1 recovery, and 1 photo graphic platform |
| | | | 6 | HC-130 | Two each staged from Hawaii, Samoa, and Ascension |
| | | | 3 | E-1B | 1 Airboss, 1 relay, and 1 Airboss/relay combination air craft |
| Primary | Crew: 16 | | 1 | CVS | USS Hornet |
| | CM: 24 | 2 | 4 | SH-3D | Two with swimmers, one for crew retrieval, and one pho tographic platform |
| | | | 2 | HC-130 | Staged from Pago Pago, Samoa |
| | | | 3 | E-1B | 1 Airboss, 1 relay and 1 Airboss/relay combination air craft |
| Contingency | | 18 | 6 | HC-130 | One each staged from Hickam AFB; Ascension; Mauritius Island; Andersen AFB, Guam; and Howard AFB, Canal Zone |

Total ship support = 6 Total aircraft support = 26 (This total is based on the recovery requirement that two IIC-130 aircraft be in support of the mission from Kindley AFB, Bermuda; Hickam AFB, Hawaii; Ascension; Mauritius Island and Howard AFB, Canal Zone; and one HC-130 aircraft from Andersen AFB, Guam and Lajes AFB, Azores.)

The two HC-130 aircraft, designated "Samoa Rescue 1" and "Samoa Rescue 2," were positioned to track the command module after it exited from S-band blackout, as well to provide pararescue capability if the command module landed uprange or downrange of the target point.

### 11.3.1 Command Module Location and Retrieval

Hornet's position was established using celestial fixes and satellite tracking methods. On the day of recovery the Hornet was stationed 5 miles north of the target point, which was located at 15 degrees 49 minutes south latitude and 165 degrees 10.0 minutes west longitude. The ship-based aircraft were deployed relative to the Hornet, and they departed station to begin the recovery activities upon receiving VHF signals from the command module.

Recovery forces first had contact with the command module on the Hornet's radar at 244:24:00 (2046 G.m.t., November 24, 1969). The rescue aircraft established S-band contact 4 minutes later, followed by VHF recovery beacon contact at 244:31:00 (2053 G.m.t.). VHF voice contact was established at 244:32:00 (2054 G.m.t.), followed by visual sighting of the command module during the descent on the main parachutes. The command module landed at 244:36:25 (2058 G.m.t.) at a point calculated by recovery forces to be 15 degrees 46.6 minutes south latitude and 165 degrees 9.0 minutes west longitude.

The command module landed in the stable I (apex up) flotation attitude and immediately went to the stable II (apex down) attitude. The uprighting system returned the command module to the stable I attitude 4 minutes 26 seconds later. After the swimmers were deployed and had installed the flotation collar, the decontamination swimmer passed flight suits and respirators to the crew, and aided the crew in entering the life raft. After the crew had been retrieved, the decontamination swimmer decontaminated the external surface of the command module.

The crew arrived aboard the Hornet at 2148 G.m.t. and entered the mobile quarantine facility 8 minutes later. The interior of the prime recovery helicopter was then decontaminated as part of the quarantine procedures.

### 11.3.2 Postretrieval Operations and Quarantine

The command module was brought aboard the Hornet at 2246 G.m.t. It was secured to the mobile quarantine facility shipboard transfer tunnel after a brief welcoming ceremony, and the lunar samples, film, and tapes were removed. The first samples to be returned were flown to Samoa, transferred to a C-141 aircraft, and flown to Houston. The second sample shipment was flown from the Hornet to Samoa, transferred to a range instrumentation aircraft, and flown to Houston.

The mobile quarantine facility was unloaded in Hawaii at 0218 G.m.t., November 29, followed shortly by the unloading of the command module. After a brief welcoming ceremony in Hawaii, the mobile quarantine facility was loaded aboard a C-141 aircraft and flown to Ellington Air Force Base, Texas. The crew arrived at the Lunar Receiving Laboratory at 1350 G.m.t. on November 29.

The command module was unloaded in Hawaii and was taken to Hickam Air Force Base for deactivation. When deactivation was completed 2-1/2 days later, the command module was flown to Ellington Air Force Base on a C-133 aircraft. The following is a chronological listing of events during the recovery and quarantine operations.

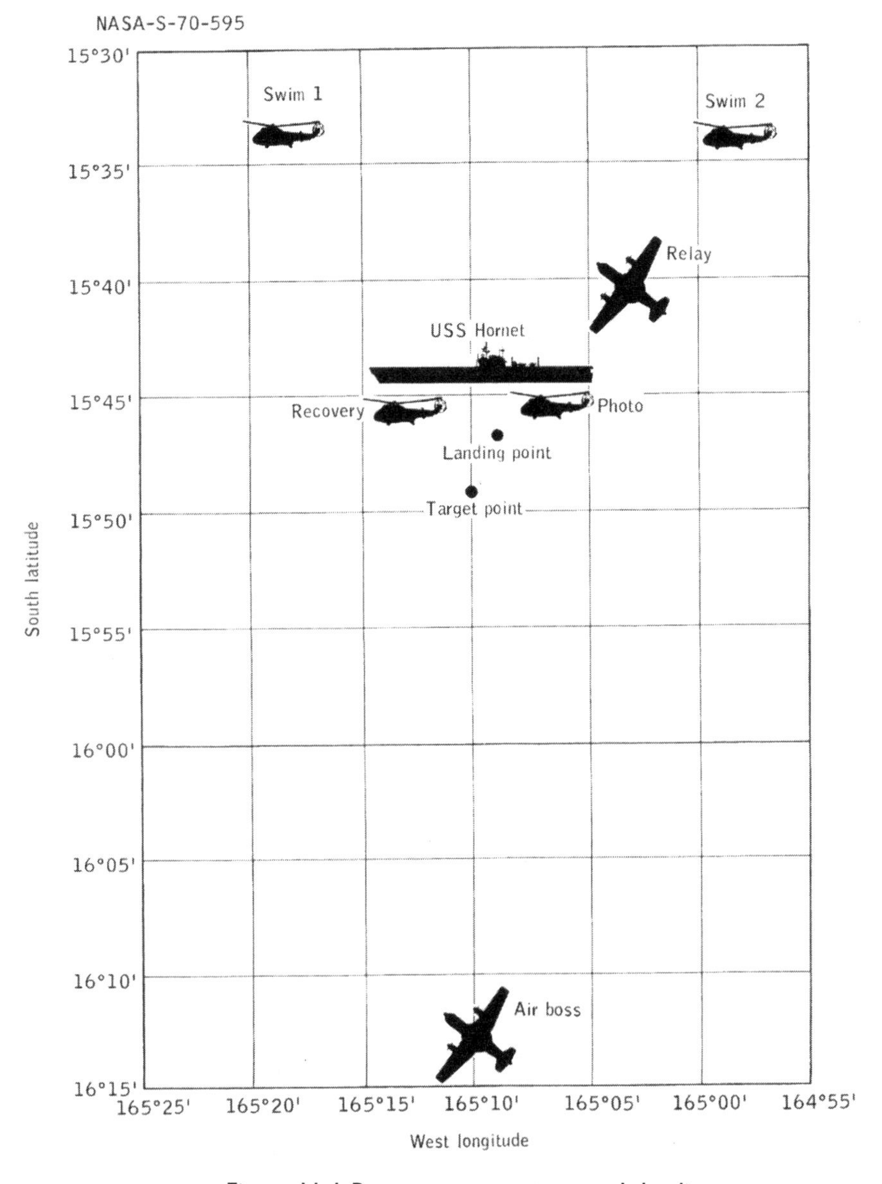


Figure 11-1 Recovery support at earth landing.

| Event | Time from lift-off, hr:min | Time, G.m.t. |
|---|---|---|
| November 24, 1969 | | |
| Radar contact by Hornet | 244:24 | 2046 |
| S-band contact by rescue aircraft | 244:28 | 2050 |
| VHF recovery beacon signals received | 244:31 | 2053 |
| VHF voice contact received by aircraft and Hornet | 244:32 | 2054 |
| Command module landed, went to stable II | 244:36 | 2058 |
| Command module uprighted to stable I | | 2103 |
| Swimmers deployed to command module | | 2108 |
| Flotation collar inflated | | 2115 |
| Command module hatch opened for respirator transfer | | 2136 |
| Command module hatch opened for crew egress | | 2140 |
| Flight crew aboard Hornet | | 2158 |
| Flight crew entered mobile quarantine facility | | 2206 |
| Command module lifted from water | | 2246 |
| | | November 25 |
| Command module secured to the mobile quarantine transfer tunnel | | 0015 |
| Command module hatch opened | | 0040 |
| Apollo lunar sample return containers 1 and 2 removed from the command module | 0152 | |
| Container 1 removed from mobile quarantine facility | | 0314 |
| Container 1 controlled temperature shipping container 1, and film flown to Samoa | | 0640 |
| Container 2 removed from mobile quarantine facility | | 0811 |
| Container 2, remainder of biological samples and film flown to Samoa | | 1130 |
| Container 1, controlled temperature shipping container 1, and film arrived in Houston | | 2045 |
| Command module hatch secured and decontaminated | | 2223 |
| Mobile quarantine facility secured after removal of transfer tunnel | | 2330 |
| | | November 26 |
| Container 2, remainder of biological samples, and film arrived in Houston | 0448 | |
| | | November 29 |
| Mobile quarantine facility and command module offloaded in Hawaii | | 0218 |
| Safing of command module pyrotechnics complete | | 0840 |
| Mobile quarantine facility arrived at Ellington AFB | | 1150 |
| Flight crew entered Lunar Receiving Laboratory | | 1350 |
| | | December 1 |
| Deactivation of the fuel and oxidizer completed | | 1415 |
| | | December 2 |
| Command module delivered to Lunar Receiving Laboratory | | 1930 |

### 11.3.3 Postrecovery Inspection

All aspects of the command module, mobile quarantine facility, and lunar sample return containers were normal except for the following discrepancies:

a. Condensation was found between the panes of the number 1 window (far left). The number 5 window (far right) had a frosty film on the outer pane and condensation on the inner pane (section 14.1.11).

b. The environmental control system hose was broken at the bulkhead connection for the center couch. The connection bracket came off the panel (section 14.1.14).

c. The camera had dislodged from its mount at landing.

d. Two whiskers on the VHF antenna did not deploy (section 14.1.12).

e. The shaped charge ring was broken but was held by the spring clips. One of these spring clips was missing.

f. Oxygen pressure was depleted during the command module water sampling operation, and no waste water or drinking water samples were taken.

# 12.0 ASSESSMENT OF MISSION OBJECTIVES

The five primary mission objectives (see reference 10) assigned the Apollo 12 mission were as follows:

a. Perform selenological inspection, survey, and sampling in a mare area

b. Deploy the Apollo lunar surface experiments package

c. Develop techniques for a point landing capability

d. Further develop man's capability to work in the lunar environment

e. Obtain photographs of candidate exploration sites.

Twelve detailed objectives, listed in table 12-I and described in reference 11, were derived from the five assigned primary objectives. The following experiments, in addition to those contained in the experiment package (see appendix A), were also assigned:

a. Lunar Field Geology (S-059)

b. Solar Wind Composition (S-080)

c. Lunar Multispectral Photography (S-158)

d. Pilot Describing Function (T-029)

e. Lunar Dust Detector (M-515).

All detailed objectives were met, with the following exceptions: objective G - Photographs of Candidate Exploration Sites, and objective M - Television Coverage. These two objectives were not completely satisfied, based on preflight planning data; the portions of these objectives not accomplished are described in the following paragraphs.

## 12.1 PHOTOGRAPHS OF CANDIDATE EXPLORATION SITES

To obtain sufficient photographic data on candidate lunar landing sites for future missions, the following coverage of lunar surface areas Lalande, Fra Mauro, and Descartes was planned:

a. 70-mm stereoscopic photography of the ground track from terminator to terminator during two passes over the three sites, with concurrent 16-mm sextant sequence photography during the first pass

b. Landmark tracking of a series of four landmarks bracketing the three sites included in the stereoscopic photography, and performed during two subsequent, successive orbits

c. 70-mm high resolution photographs using a 500-mm lens, and additional high resolution oblique photography.

The first 70-mm stereoscopic photography pass, the concurrent 16-mm sextant sequence photography, and the first landmark tracking series were accomplished. The necessity to repeat high resolution photography did not provide sufficient time to complete both the second stereoscopic photography pass and the second landmark tracking series. A real-time decision assigning higher priority to landmark tracking therefore allowed tracking of the two landmarks associated with Fra Mauro and Descartes and completion of about one-fourth of the second stereoscopic photography pass.

Because of a crew error in site identification, the first high resolution photographs were taken of the Herschel area instead of Lalande. However, a substitute target to the south of Lalande, assigned in realtime, was subsequently photographed. A first attempt to obtain high resolution photographs of Fra Mauro and Descartes was unsuccessful because of a camera malfunction (see section 14.3.7). However, on a second attempt, photographs were obtained of Fra Mauro and an area slightly east of the Descartes target area, and high resolution oblique photography was also accomplished.

In summary, all mandatory requirements were satisfied with the exception of about three-fourths of the second stereoscopic photography pass and tracking of two landmarks of the second landmark tracking series. All highly desirable requirements were satisfied except for the planned high resolution photography of Descartes. Photographic requirements of this objective not accomplished are planned for future Apollo missions, although the candidate sites selected for photography might differ.

## 12.2 TELEVISION COVERAGE

No specific priority was assigned to the objective of general television coverage because television requirements were to be satisfied as a part of other objectives. Television requirements consisted of obtaining coverage of:

a. A crewman descending to the lunar surface

b. An external view of the landed lunar module

c. The lunar surface in the general vicinity of the lunar module

d. Panoramic coverage of distant terrain features

e. A crewman during extravehicular activity.

Coverage was obtained only of a crewman descending to the lunar surface. The other coverage was not obtained because the camera was damaged immediately after it was removed from its stowage compartment (see section 14.3.1). This objective is planned again for Apollo 13.

### TABLE 12-I.- DETAILED OBJECTIVES AND EXPERIMENTS

| Description | | Completed |
|---|---|---|
| A | Contingency sample collection | Yes |
| B | Lunar surface extravehicular operations | Yes |
| C | Portable life support system recharge | Yes |
| F | Selected sample collection | Yes |
| G | Photographs of candidate exploration sites | Partial |
| H | Lunar surface characteristics | Yes |
| I | Lunar environment visibility | Yes |
| J | Landed lunar module location | Yes |
| L | Photographic coverage | Yes |
| M | Television coverage | Partial |
| N | Surveyor III investigation | Yes |
| 0 | Selenodetic reference point update | Yes |
| ALSEP I | Apollo lunar surface experiments package | Yes |
| S-059 | Lunar field geology | Yes |
| 5-080 | Solar wind composition | Yes |
| S-158 | Lunar multispectral photography | Yes |
| T-029 | Pilot describing function | Yes |
| M-515 | Lunar dust detector | Yes |

# 13.0 LAUNCH VEHICLE SUMMARY

The trajectory parameters of the AS-507 launch vehicle from launch to translunar injection were close to expected values. The vehicle was launched on an azimuth 90 degrees east of north. A roll maneuver was initiated at 12.8 seconds to place the vehicle on a flight azimuth of 72.029 degrees east of north.

Following lunar module ejection, the vehicle attempted a slingshot maneuver to achieve a heliocentric orbit. However, the vehicle's closest approach of 3082 miles above the lunar surface did not provide sufficient energy to escape the earth-moon system. Even though the slingshot maneuver was not achieved as planned, the fundamental objective of not impacting the spacecraft, the earth, or the moon was achieved. The vehicle did not achieve a heliocentric orbit because the computed time for auxiliary propulsion ullage firing was based on the telemetered state vector, which was within the 3-sigma limit but was in excess of the 13.1 ft/sec slingshot window velocity.

In the S-IVB stage, the oxygen/hydrogen burner satisfactorily achieved tank repressurization for restart. However, burner shutdown did not occur at the programmed time due to an intermittent electrical open circuit, and this resulted in a suspected burnthrough of the burner. Subsequent engine restart conditions were within specified limits, and the restart at full-open propellant utilization valve position was successful. The electrical systems performed satisfactorily throughout all phases of flight except during the S-IVB restart preparations. During this time, the S-IVB stage electrical systems did not respond properly to burner liquid oxygen shutdown valve "close" and telemetry calibrate "on" commands from the S-IVB switch selector. All hydraulic systems performed satisfactorily, and all parameters were within limits, although the return fluid temperature of one S-IC actuator rose unexpectedly at 100 seconds.

This Apollo/Saturn vehicle was the first to be launched in inclement weather, and two distinct lightning strikes occurred (reference 12). However, the structural loads and dynamic environments experienced by the vehicle were well within the structural capability.

Low-level oscillations, similar to those of previous flights, were evident during each stage firing but caused no problems. The S-II stage experienced four new periods of 16-hertz oscillations, which apparently result from the inherent characteristics of the present S-II stage configuration; however, engine performance was not affected.

# 14.0 ANOMALY SUMMARY

This section contains a discussion of the significant problems or discrepancies noted during the Apollo 12 mission. Anomalies in the operation of experiment equipment after deployment will be published in a separate anomaly report.

## 14.1 COMMAND AND SERVICE MODULES

### 14.1.1 Intermittent Display and Keyboard Assembly

The crew reported several intermittent, all-"8's" displays on the main display and keyboard assembly approximately 1-1/2 hours before launch, but no display malfunction occurred in flight. The display segments are illuminated by applying 250 V AC through the contacts of miniature relays, as shown in figure 14-1. When a segment is off, it is grounded through a resistor and the normally closed contacts of a relay to avoid residual illumination. The normally closed contacts of all relays are tied together; consequently, a short across the contacts of any one relay will apply the voltage to all segments of each display. The effect of the short in conjunction with the common discharge path is shown in figure 14-1 for a typical character and one sign. A short across the relay contacts will affect only the display function of the unit.

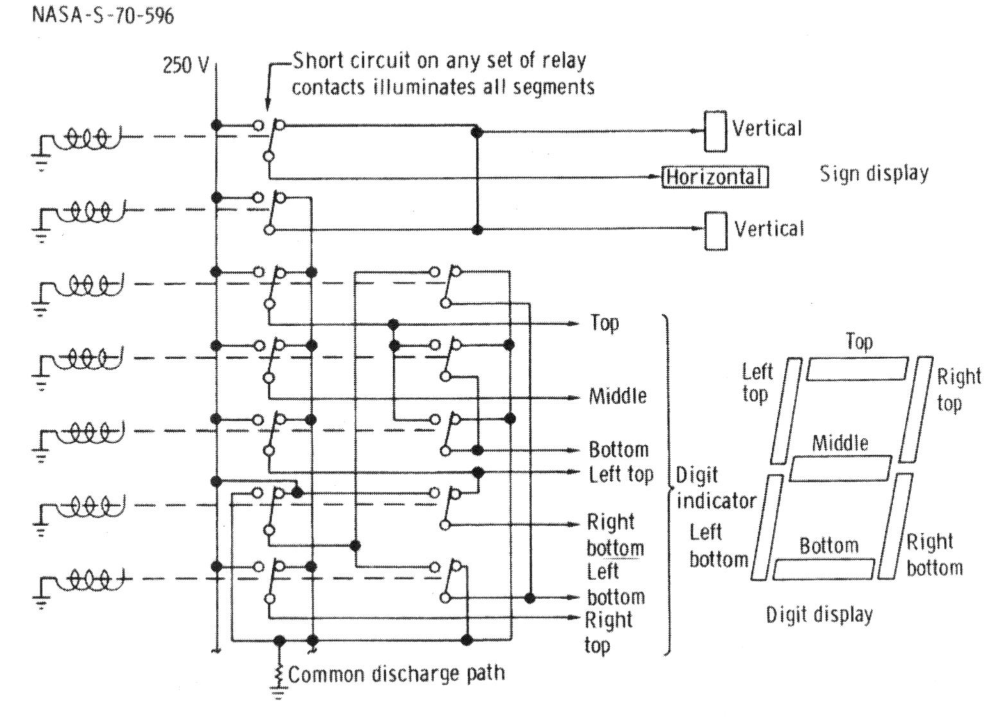


Figure 14-1 Simplified schematic diagram of relay matrix.

Failure analyses performed after four previous identical failures on other units showed that contamination was present in a relay which could have caused the all-"8's" display. As a result, the fabrication, process has been improved through the use of laminar-flow clean rooms to minimize contamination. A 100-percent vibration screening procedure was initiated at the part level with automatic detection of any actuation faults. After assembly, each display keyboard is vibrated during actual operation and visually observed for fault detection. However, improved fabrication techniques and test procedures can not eliminate the possibility of contamination; consequently, a malfunction procedure has been devised to remove a shorted condition through the actuation of all relays.

This anomaly is closed.

### 14.1.2 Hydrogen Tank Leakage

During cryogenic loading about 51 hours before launch, the heat leak of hydrogen tank 2 was unacceptable. Visual checks showed a thick layer of frost on the tank exterior, verifying an inadequate vacuum in the insulating annulus. The tank was removed and replaced. A failure analysis performed before launch identified the cause of the vacuum loss as an incomplete bond in the stainless steel/titanium bimetal joint, which permitted hydrogen to leak from the inner tank into the annulus (fig. 14-2). The bimetallic joint provides a seal between the two metals, which are not compatible for welding to each other. The joint is made from a billet such that the two metals are extruded together and machined. The machined fitting is welded in place, as shown in figure 14-2.

Improper inspection of the bimetallic joint during manufacture has allowed voids between the metal surfaces to pass unnoticed. The failed joint was manufactured in lot 3B, and lot 3A was also suspected as having poor quality joints. There are only four other tanks from these two lots remaining in the program, and these tanks have been recalled for replacement of the questionable bimetallic joints.

This anomaly is closed.

### 14.1.3 Electrical Potential Discharges

The spacecraft and launch vehicle were involved in two lightning discharge during the first minute of flight. The first, at 36.5 seconds after lift-off, was from the clouds to earth through the vehicle. The second discharge involving the vehicle occurred at 52 seconds and was from cloud to cloud. The two discharges were distinctly recorded by groundbased instrumentation.

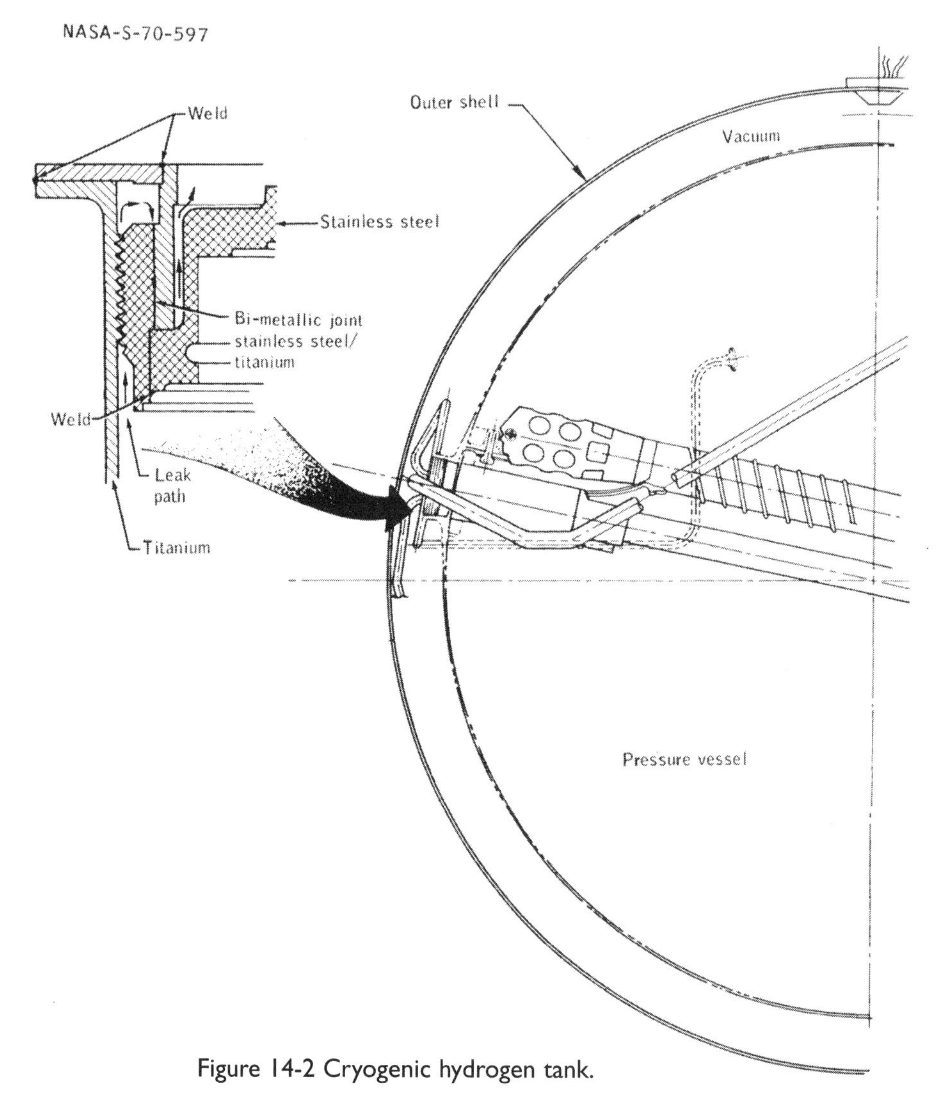


Figure 14-2 Cryogenic hydrogen tank.

The discharge at 36.5 seconds disconnected the fuel cells from the spacecraft buses and damaged nine instrumentation measurements. The discharge at 52 seconds caused loss of reference in the spacecraft inertial platform. Both discharges caused a temporary interruption of spacecraft communications. Many other effects were noted on instrumentation data from the launch vehicle, which apparently sustained no permanent damage from the discharges.

A complete analysis of the lightning incidents and the associated phenomena is presented in a special report (reference 12). This report attributes the lightning to the presence of the vehicle, as it passed through electric fields sufficient in intensity and energy to trigger each discharge.

Instrumentation loss.- The only permanent effect on the spacecraft was the loss of nine measurements at the first discharge. Of these nine, four were service module outer surface temperature sensors, four were reaction control system propellant quantity measurements, and one was a temperature measurement on the nuclear particle analyzer. All of the failed measurements are located on the service module near the interface of the command and service modules.

The service module outer surface temperature measurements use a chromel-constantan thermocouple and a reference junction. The reference junction is a bridge made up of three resistors and a temperature sensitive diode (fig. 14-3). The resistors normally operate at about 0.020 ampere and will open in the region of 0.100 ampere. An open bridge resistor would drive the signal output off-scale high or low depending upon which resistor fails.

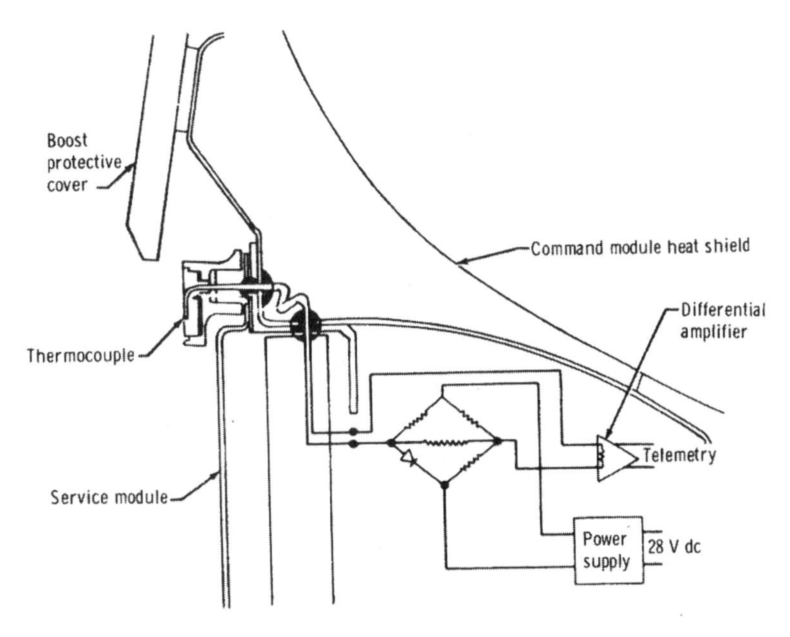


Figure 14-3 Simplified schematic and location of a typical outer skin temperature sensor.

It is probable that the nuclear particle analyzer temperature failed as a result of burning out a zone box resistor in a manner similar to the outer surface temperature sensor failures.

The reaction control propellant quantity measurements use semiconductor strain gages on a pressure-sensitive diaphragm (fig. 14-4). The semiconductors are a thin film type, and excessive current would probably damage their capability to operate as pressure-sensitive resistors. An alternate possibility is that the Zener diode, used to regulate the 14-volt supply to 6.4 volts, was burned out. Loss of this diode would explain the instrumentation symptom, which in all four cases was full-scale and unchanging.

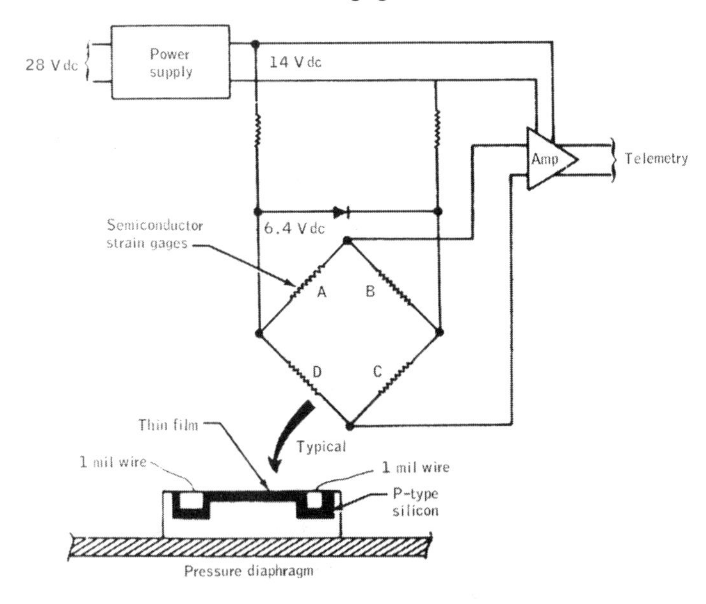


Figure 14-4 Propellant quantity transducer schematic.

Fuel cell disconnect.- At the time of the first lightning discharge, the fuel cells were automatically removed from the spacecraft buses with the resultant alarms normally associated with total fuel cell disconnection.

The voltage transient that was induced on the battery relay bus by the static discharge exceeded the current rate-of-change characteristics of the silicon controlled rectifiers in the fuel cell overload sensors and disconnected the fuel cells from the bus (fig. 14-5). As a result, the main bus loads of 75 amperes were being supplied totally by entry batteries A and B, and the main bus voltages dropped momentarily to approximately 18 or 19 volts, but recovered to 23 or 24 volts within a few milliseconds. The low dc voltage on the main buses resulted in the illumination of undervoltage warning lights, a drop out of the signal conditioning equipment, and a lower voltage input to the inverters. The momentary low-voltage input to the inverters resulted in a low output voltage which tripped the ac undervoltage sensor causing the ac bus 1 fail light to illuminate. The transient that disconnected the fuel cells from the buses also caused the silicon controlled rectifier in the overload circuits to indicate an ac overload. At 2 minutes 22 seconds into the flight, the crew restored fuel cell power to the buses. All bus voltages remained normal throughout the remainder of the flight.

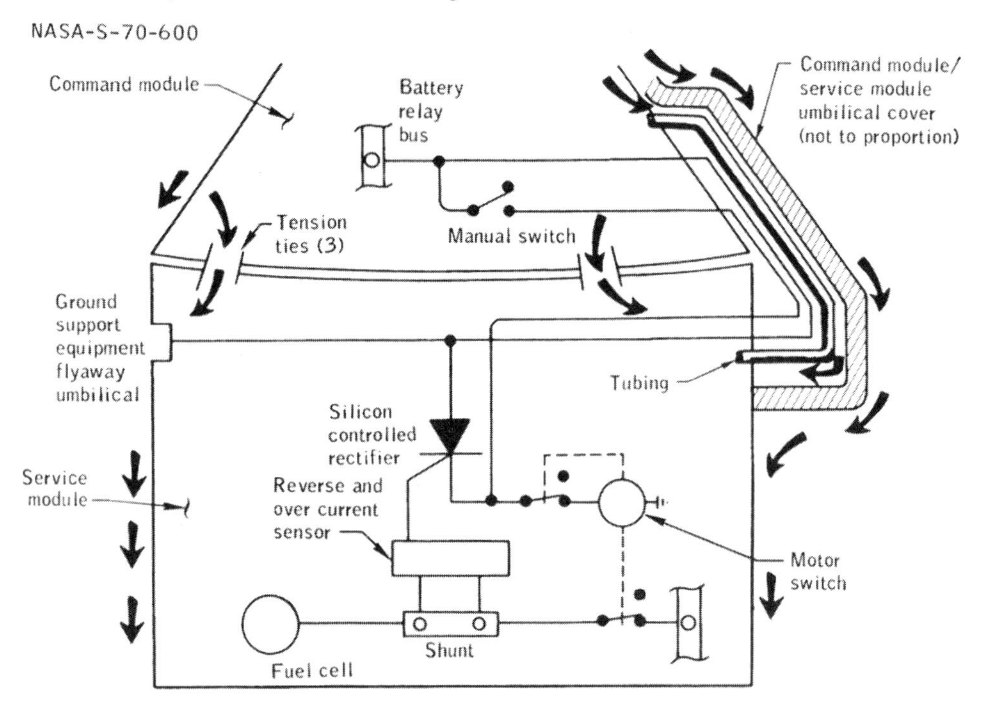


Figure 14-5 Fuel cell disconnection circuitry.

Loss of inertial platform reference.- A loss in reference for the inertial platform at the second discharge was most likely caused by the setting of high-order bits in the coupling display unit by the discharge transients introduced between signal ground and structural ground. If this condition occurs and causes the Z-axis (yaw) coupling display unit (middle gimbal) readout to exceed 85 degrees, the computer will down-mode the platform to coarse align. When the coupling display unit is driving at high speed to null the noise-induced error and the coarse-align loop is energized, the servo loop from the coupling display unit to the platform becomes unstable and drives the platform in the manner observed. A change to the computer programing to inhibit the computer mode-switching logic during the launch phase has been implemented for Apollo 13.

Complete protection of the spacecraft from the effects of lightning is not considered practical at this stage of the program. The inherent temporary effects associated with solid state circuitry and the reasonable degree of safety in other circuits warrants the low risk of triggering lightning if potentially hazardous electric fields are avoided.

The following launch restrictions have been imposed for future missions to greatly minimize the possibility of triggering lightning.

a. No launch when flight will go through cumulonimbus (thunderstorm) cloud formation. In addition, no launch if flight will be within 5 miles of thunderstorm clouds or within 3 miles of an associated anvil.

b. Do not launch through cold-front or squall-line clouds which extend above 10 000 feet.

c. Do not launch through middle cloud layers 6000 feet or greater in depth where the freeze level is in the clouds.

d. Do not launch through cumulus clouds with tops at 10 000 feet or higher.

This report reflects the combined efforts of the investigating teams at the Manned Spacecraft Center, the Kennedy Spacecraft Center, and the Marshall Space Flight Center.

This anomaly is closed.

### 14.1.4 Open Stabilization and Control System Circuit Breaker

During systems checks after earth orbit insertion, circuit breaker 23 for stabilization and control logic bus 3 and 4 on panel 8 was found in the open position (fig. 14-6). A crewman closed the circuit breaker and it remained closed throughout the rest of the mission. Complete electrical and mechanical tests were performed and the results were normal. The circuit breaker and associated circuitry showed no cause for the breaker to have opened either because of launch vibrations or an electrical fault.

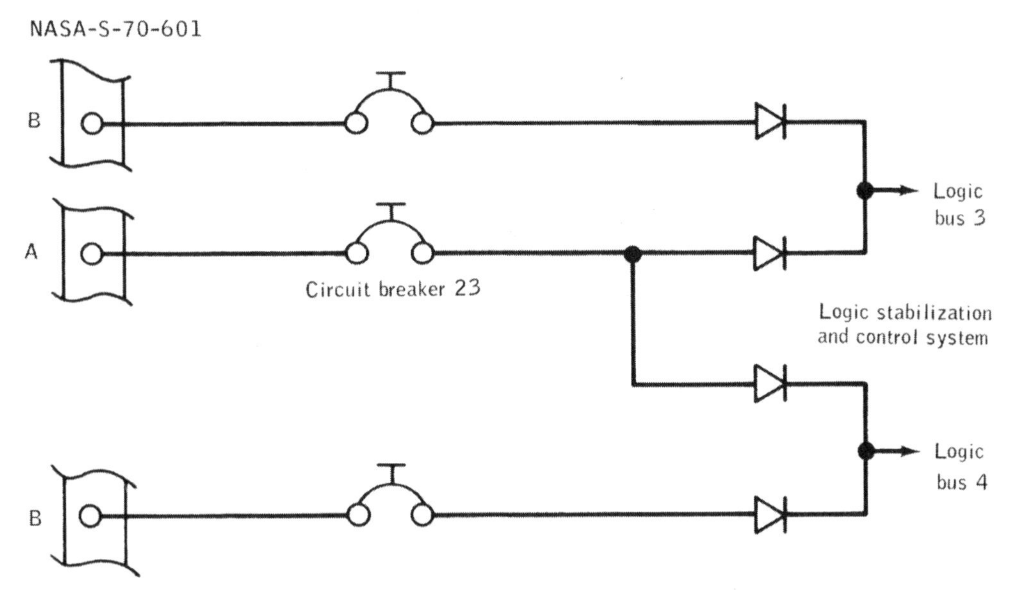


Figure 14-6 Stablization and control circuit breaker schematic.

As shown in the figure, the breaker was supplying power in parallel with two other breakers which did not open. This fact plus no abnormalities indicate that the breaker was probably not set during the prelaunch switch and circuit breaker positioning checks. These breakers are not specifically verified to be in proper position.

This anomaly is closed.

### 14.1.5 Inadvertent Helium Isolation Valve Closure

The crew reported that two isolation valves had inadvertently closed during the command and service module/S-IVB separation sequence. The quad A secondary propellant isolation valve and the quad B number 1 helium isolation valve closed. The crew reopened the valves according to preplanned procedures, and no further problems were experienced. This same phenomenon occurred during the Apollo 9 and 11 missions for propellant isolation valves, but the closing of the helium isolation valve was the first noted inflight occurrence. The failure investigation test programs for Apollo 9 and 11 led to the conclusion that valve closures can be expected because of the separation shock levels produced by the pyrotechnics, and, that these closures are not detrimental to the valves.

This is the first instance that a helium isolation valve has closed, and some differences exist between the helium and propellant isolation valves. The helium valve requires a slightly lower force to close, since the poppet mass is slightly higher and the seat configuration is different.

An analysis of propagation and intensity of the shock at S-IVB separation indicates intensities of 45g to 275g, random in direction and lasting 1 to 3 milliseconds. The valves are qualified for 7g shocks of 11 milliseconds duration in all six direction. Therefore, it is possible that the valves could close when subjected to the S-IVB separation shock.

Component testing was conducted on the propellant isolation valve to establish the sensitivity threshold and has shown that shocks of 80g to 130g with durations of 11 to 1 milliseconds, respectively, can cause an open valve to close. Further tests showed that these valves, as well as a valve that was repeatedly closed with a 280g shock for 3 milliseconds, were in no way damaged or degraded by the shocks. Flight experience also indicated no adverse effects due to the closures.

The helium isolation valve was not tested, but an analytical evaluation indicates that the valve will change position at lower g forces than those required to close the propellant valves, primarily because of the higher poppet mass. The orientation of the valve and/or possible attenuation may explain the smaller frequency of occurrence compared to the closing of the propellant valves. Tests have indicated that the minimum shock on the helium valves, in the direction of poppet movement, is about 45g for 1 to 3 milliseconds. The maximum comparable shock on the propellant valves is estimated to be 270g for 1 to 3 milliseconds.

Analysis of the helium isolation valve indicates that, because of the valve seat construction and the lower level of shock, no functional degradation can occur as a result of the separation shock. Procedures will be maintained to verify the position of these valves after separation from the S-IVB.

This anomaly is closed.

### 14.1.6 S-band Signal Strength Variations

Operation of the S-band high gain antenna in the narrow beam mode resulted in a decrease of approximately 10 to 12 dB in both uplink and downlink signal strength on several occasions. Illustrations of the first and other unexpected signal-strength variations are shown in figure 14-7. The first decrease occurred in lunar orbit revolution 1.

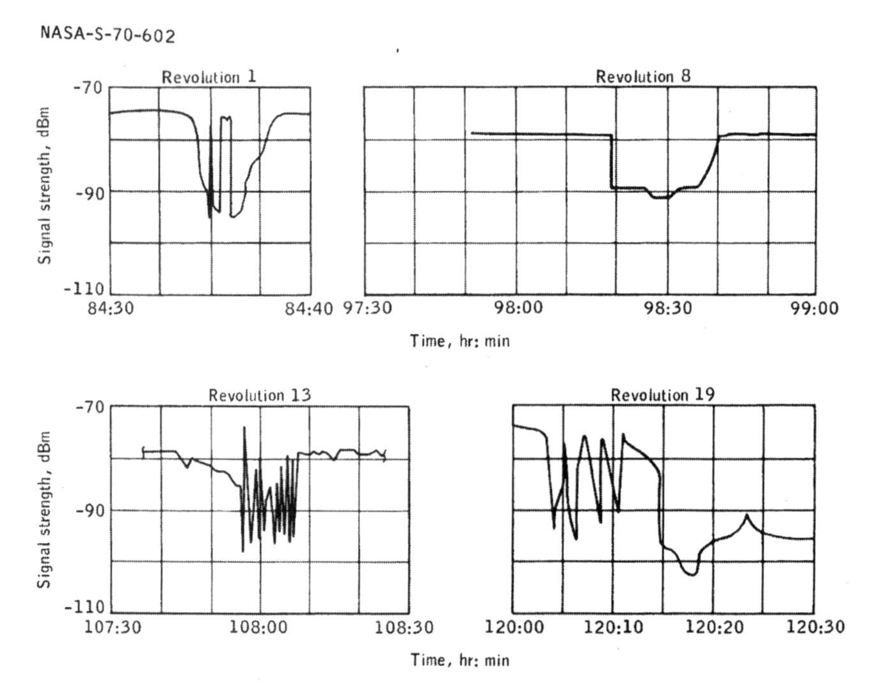


Figure 14-7 Typical high gain antenna uplink signal strength during abnormal operations.

Two special tests were conducted during transearth coast with the spacecraft in attitude hold to isolate the malfunction. The sun angle was within approximately a 12-degree cone about the minus X axis to induce thermal stress on the antenna. In both tests, the narrow-beam and reacquisition modes were maintained until fluctuations in the uplink and downlink signal strengths were observed. When a dropout appeared during the first test, the mode was changed to wide beam and the signal strength became normal. The second test included acquisition in the wide beam mode after signal-strength fluctuations had been observed in narrow beam, and normal signal levels were restored after acquisition.

Based on antenna-related data during lunar orbit and from the special tests, the problem can be summarized as follows:

a. Signal strength was reduced at about the same magnitude in both the uplink and downlink signals while in narrow beam

b. The magnitudes of the reductions were generally from 10 to 12 dB and usually of a gradual change at first

c. The malfunction occurred only in automatic and auto-reacquisition narrow-beam modes

d. A normal signal could be restored by switching to the manual mode and aligning the antenna to earth

e. Switching between primary and secondary electronics caused no change in operation

f. The malfunction occurred after a period of proper tracking in the narrow-beam mode, but not during acquisition

g. After occurrence of this malfunction, operation at times returned to normal without switching by the crew

h. The malfunction occurred in regions near both the center and the scan limits of the antenna

i. Three tracking stations reported that very large 50-hertz and smaller 400-hertz spikes appeared on the dynamic-phase error displays when signal-strength reductions existed.

Laboratory tests, conducted for further analysis of the last item, verified that spikes in the dynamic phase error response of the ground station receiver could be generated by introducing square wave modulation on the up- or downlink at the spacecraft terminal. Since these tests were performed with a bench modulator and not the actual flight hardware, it could not be definitely determined if the modulation was introduced on the uplink or downlink. The normal operation of the antenna when not boresighted will introduce square-wave modulations of the uplink signal because of the lobing sequence. If the tracking stations were observing this downlink modulation, then the cause is a malfunction of the antenna stripline units.

An analysis of antenna feeds consisted of eliminating one dish from the narrow beam array. This analysis was first accomplished by considering the case of no contribution from one dish and then determining the contribution from one dish 180 degrees out of phase. With no contribution from one dish, the boresight shift was slightly over 1 degree and the accompanying gain loss was 1.5 dB, which was much less than the 10 or 12 dB loss recorded during flight. It is apparent that the antenna will track with one dish inoperative and with the previously mentioned boresight shift and gain losses. One dish having a phase error of 180 degrees will tend to produce boresight shifts of greater than 5 degrees, which correspond to gain reductions of approximately 10 dB. Creation of such a phase shift in the feeds or lines prior to the comparator is very remote. Phase shifts of this order are more likely to have been produced in the stripline units.

There is a total of four stripline units with one contained in each of the following antenna components: narrow-beam comparator, transfer switch, and dual diplexer, as shown in figure 14-8. Based on the inflight tests, the wide beam comparator has been eliminated as a cause of the anomaly. Also, investigation of the circuitry and correlation of data has ruled out the transfer switch as being the anomaly cause. Therefore, the malfunction could only have been in the narrow beam comparator or the dual diplexer.

The narrow beam comparator combines the patterns of four dish antennas to provide the sum and difference patterns which provide the angle pointing information. Two malfunctions that could produce boresight shifts have been identified in the narrow beam comparator. Under normal operating conditions, the lobing switches function as digital phase shifters and provide either a zero or 180-degree phase shift. If a diode fault occurs that changes the phase or amplitude characteristics of either switch, tracking errors can be produced. The opening of one set of diodes would have to be intermittent to produce the observed flight anomaly, thus suggesting the presence of temperature or pressure sensitive connections in the traces that connect the diode switches. Another diode fault which can occur is a loss of the drive voltage to one of the lobing switches. In this case, the switch will provide a constant phase shift. The multiplexed difference signal for the case when the phase shift of switch 1 (fig. 14-8) is constant at 180 degrees results in unsymmetrical lobing. The antenna, in this case, will seek those pointing angles that make the elevation and azimuth angles equal in magnitude, thus suggesting that the resultant tracking error could be large and not repeatable. This condition would give the observed antenna performance characteristics. Therefore, a malfunction in the diode wiring or circuit connections is suggested. The intermittence associated with this malfunction could be explained by a temperature sensitive circuit connection (solder crack or wire break).

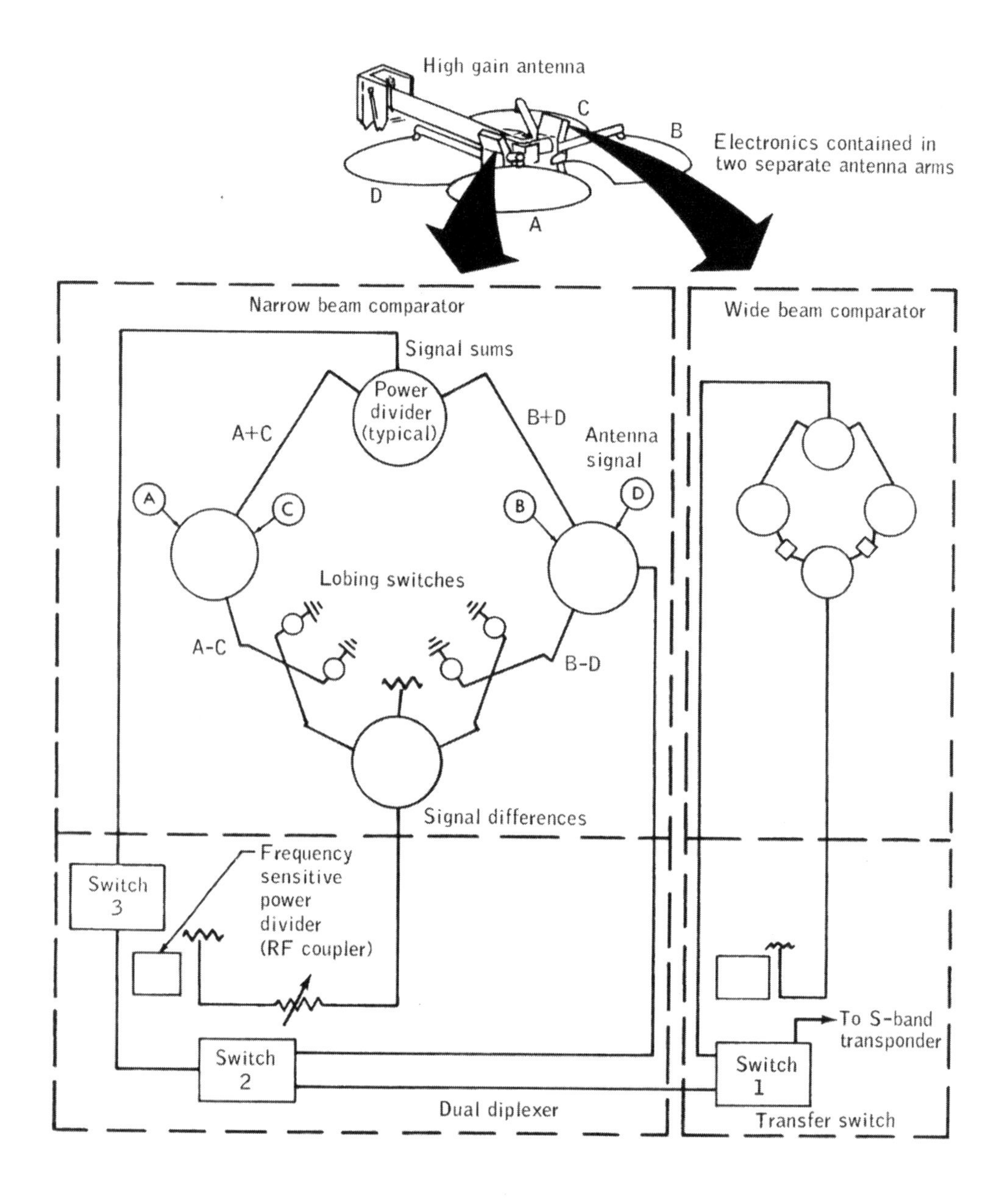


Figure 14-8 S-band high gain antenna electronics.

The major components of the dual diplexer are switches 2 and 3 and the frequency-selective power divider. The power divider is the most susceptible component for generating tracking errors. If the difference signal is attenuated by a high impedance feed-through or by incorrect phasing between the sum and difference signals, the slope of the antenna index-of-modulation curve is reduced. This decrease, in effect, reduces the total loop gain and results in an overdamped tracking system. In this case, large tracking errors would result and an antenna drift would be observed; these were the observed symptoms. Attenuation of the multiplexed difference signal can result from a trace crack or intermittent feed-through between the narrow beam comparator and the dual diplexer. Both types of failures tend to be temperature sensitive.

Malfunctions in the dual diplexer or narrow beam comparator are considered to have the highest probability as causes of the anomaly. New phase-III striplines, which should eliminate the problem, will be used on Apollo 13 and subsequent spacecraft.

This anomaly is closed.

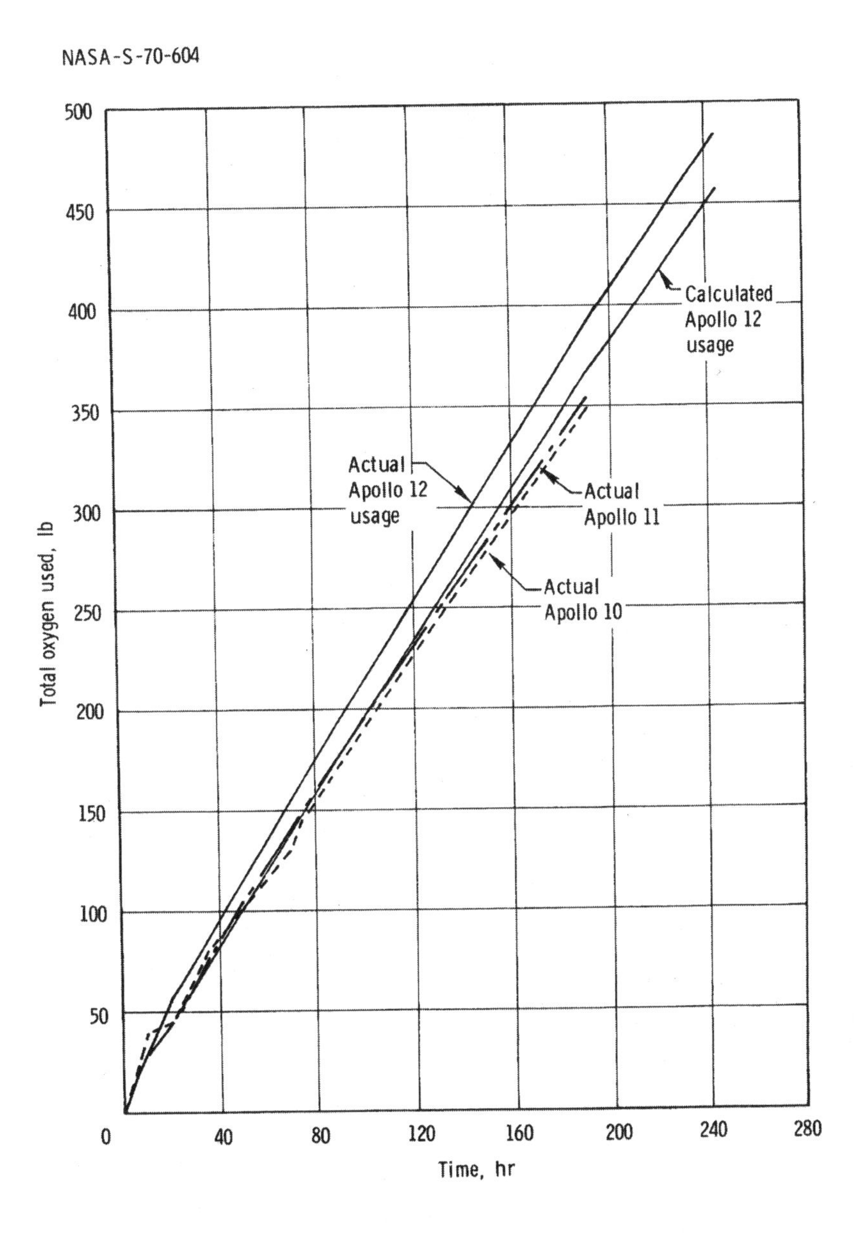


Figure 14-9 Comparison of oxygen consumption for Apollo 10, 11, and 12.

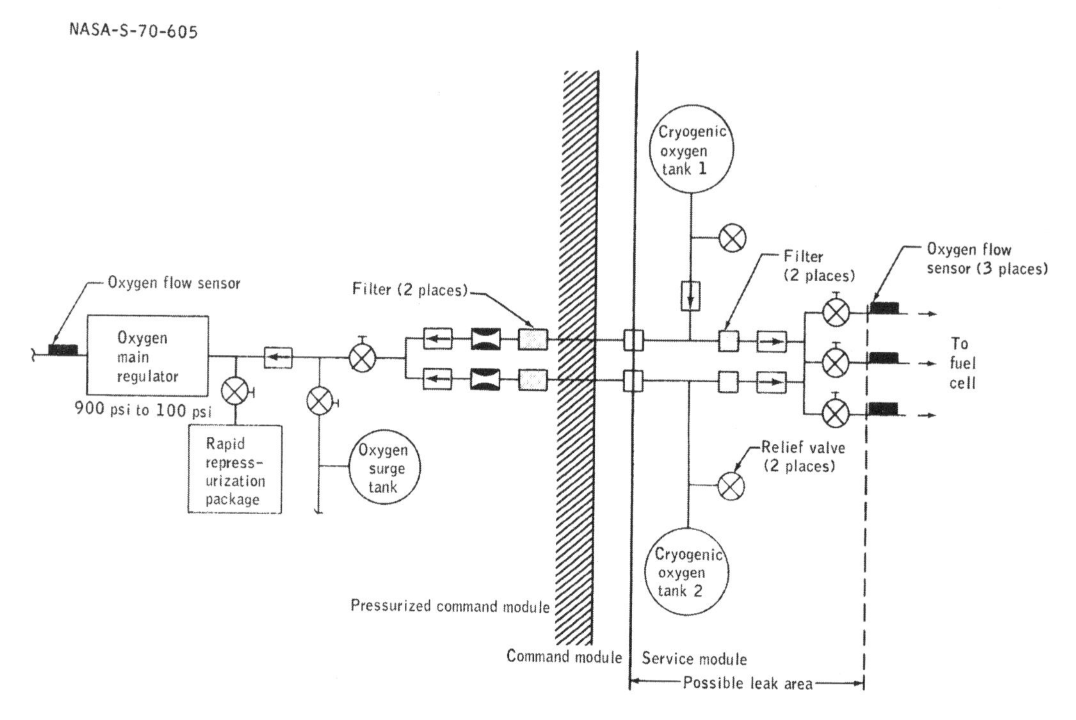


Figure 14-10 Schematic of 900 psi oxygen system.

### 14.1.7 Discrepancy in Indicated Oxygen Usage

At the end of the mission there was a discrepancy of approximately 27 pounds of oxygen between the measured total cryogenic oxygen usage and the calculated combined environmental control system and fuel cell oxygen usage, as shown in figure 14-9.

Fuel cell oxygen usage was calculated from the produced electrical current and then verified by comparison with hydrogen consumption data. Environmental control system usage is measured on a flowmeter and compared with calculated usage based on purge rates, cabin leakage rates, metabolic consumption and urine dump losses. Cabin leak rates are determined by ground tests in conjunction with flight pressure decay rates. Purge rates are calculated based on ground tests and known times for purges. Oxygen losses during urine dump operations can only be estimated. Since no excessive flow was detected downstream of the flowmeter, the source of any command module environmental control system leakage is therefore limited to the 900-psi system upstream of the meter. Figure 14-10 shows the 900psi oxygen system and that portion of the system outside the command module that could have leaked.

Postflight leak tests were conducted on the command module 900-psi system, including all check valves. These tests indicated that system leakages were within specification limits. It is therefore concluded that the 27 pounds of oxygen must have leaked from those portions of the 900psi system within the service module. Tests of these systems prior to flight are considered adequate, and no corrective action is required.

This anomaly is closed.

### 14.1.8 Material Near Service Module/Adapter Interface

The crew reported a curved piece of material about 3 feet long in the area of the service module/adapter interface. The construction of the debris catchers, charge holders, and spacecraft structure in the vicinity of the service module/adapter separation plane joint has been reviewed, and these items have been compared with pieces of material seen in Apollo 9, 10, and 12 photographs of the same area. Positive identification of the material was not possible because of the small sizes of the pieces. Photographs of Apollo 10 show two objects about 60 degrees apart near this separation plane. The crew of Apollo 12 viewed the Apollo 10 photographs and stated that the objects were similar to what they had seen during Apollo 12. Because similar pieces of material have existed on other flights without any degradation to spacecraft operation and since it is believed that no failures could occur as a result of these loose pieces, no hardware changes need be made.

This anomaly is closed.

### 14.1.9 Zero Optics Mode Fluctuations

The computer register which contains the angular position of the optics shaft was observed to fluctuate as much as 0.7 degree when the system was placed in the zero optics mode. The crew reported that the shaft mechanical readout on the optics also reflected the fluctuation.

A number of components in the optics drive servomechanism (fig. 14-11) are used only in the zero optics mode. The optical unit and the power-and-servo assembly were removed from the spacecraft, and the servo assembly was subjected to thorough testing. The flight symptoms, however, could not be reproduced. Because of extensive sea-water corrosion, the optical unit could not be tested, but an analysis and testing of a similar unit demonstrated the cause of the zero optics anomaly to be within the power-and-servo assembly. The flight assembly was installed in a working system and has operated properly under a variety of thermal conditions. The modules associated with the optics servo were also thermally cycled in an oven, operated in a vacuum, and subjected to acceptance test vibration levels with no degradation of their performance. The modules were depotted and examined, but no cause of the anomaly could be isolated.

Analysis of the circuitry involved in the zero optics mode has isolated the problem to either a relay module, a two-speed switch module, or the motor drive amplifier module. Of these, the motor drive amplifier module is the most likely cause of the anomaly observed, since it contains the only active signal-shaping network. The inflight symptoms have been reproduced on a breadboard mockup of the system by introducing a noise of from 600 to 800 millivolts into the in-phase carrier. A number of component malfunctions and shielding failures could combine to provide the avenue for introducing this level of noise. However, no evidence of a generic problem or design deficiency has been isolated; nor has system performance or component operation been affected. Therefore, no system changes are planned.

This anomaly is closed.

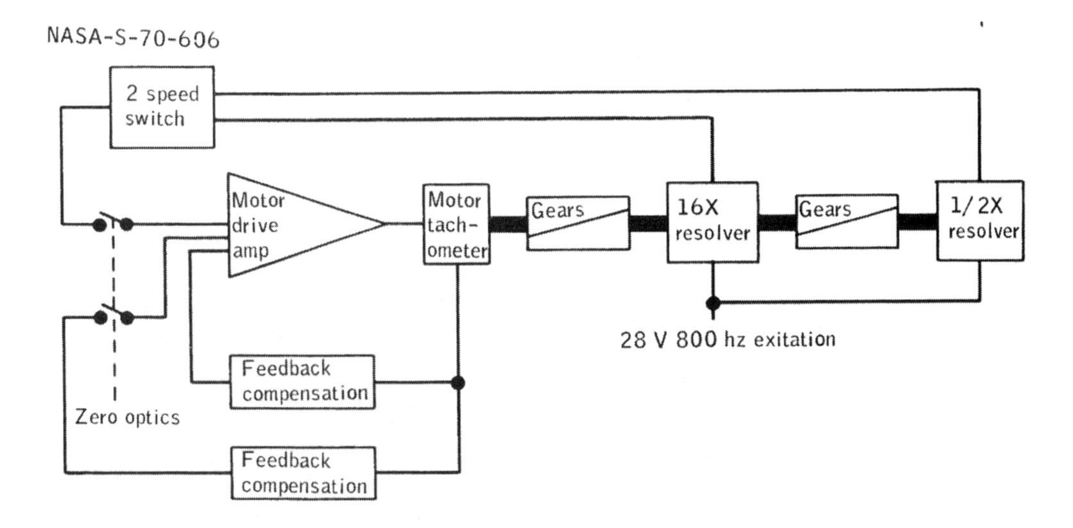


Figure 14-11 Zero optics mode circuitry.

### 14.1.10 Urine Filter clogging

By about 215 hours, the crew reported that both urine filters had clogged and that the urine overboard dump system was being operated without a filter. The inline filter (fig. 14-12) clogged the day after the Commander and the Lunar Module Pilot returned to the command module from the lunar surface activity (day 7). The filter was then replaced by a spare unit which also clogged 2 days later. The urine dumping system operated satisfactorily without a filter for the remainder of the mission (approximately 30 hours).

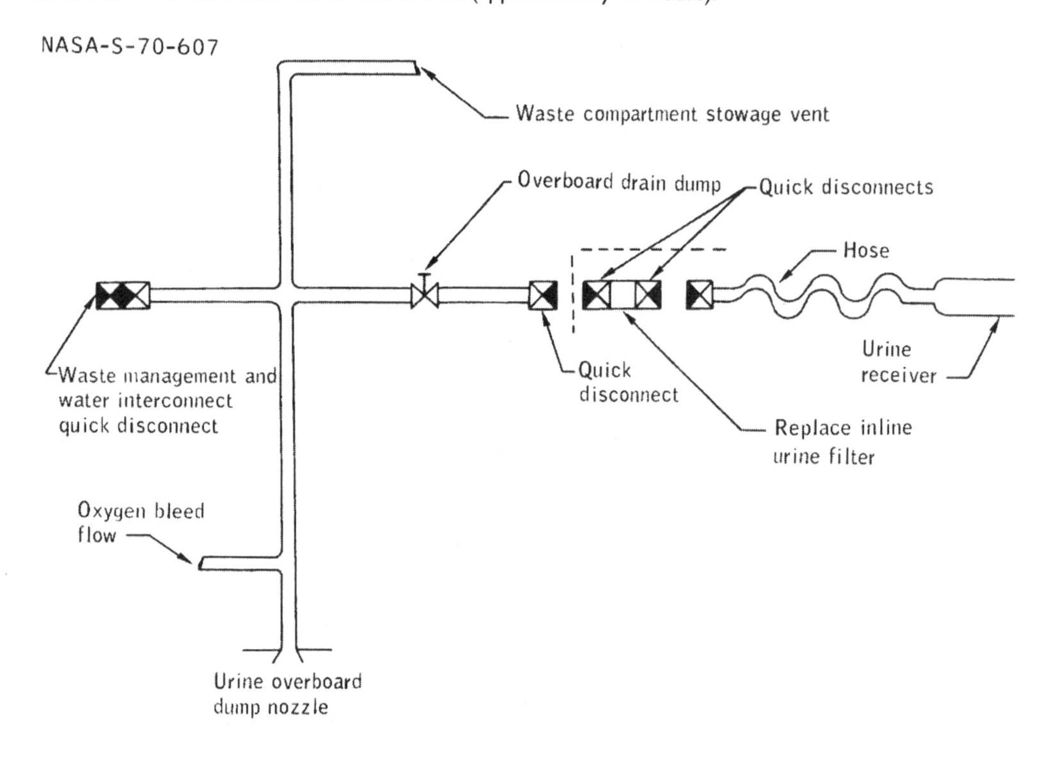


Figure 14-12 Urine dump system flow schematic.

Postflight test of both filters indicated that the clogging was primarily due to urine solids. One filter was removed from the spacecraft while in quarantine and decontaminated by autoclaving at the Lunar Receiving Laboratory. Subsequent flow and pressure drop tests were normal with the clogging material apparently removed by the autoclaving.

An analysis of the flushing water residue revealed urine solids and a small trace of lubricant but no lunar material.

The other filter was not subjected to the autoclaving process. Initial tests showed the filter was clogged, allowing only about 20 percent of normal flow. Subsequent testing showed the contamination was soluble and as the testing continued, the flow through the filter returned to normal. Analysis indicated the major contamination was urine solids. Only one small particle of lunar dust was detected in the filter.

Urine was stored in the collection device during rest periods and was to be dumped later so as to avoid perturbations to spacecraft dynamics. Previous tests have showed that storage of urine can promote formation of solids sufficient in size and quantity to plug the filter.

To minimize the problem, urine storage on future missions will be limited to critical mission time. An additional spare filter also will be stowed as a further measure.

This anomaly is closed.

### 14.1.11 Window Contamination

The hatch, left-hand side, and both rendezvous windows of the command module had considerable amounts of contamination appearing as vertical streaks on the exterior surfaces. Before flight, gaps in the boost protective cover were noted in the hard-to-soft transition region over the left rendezvous window (fig. 14-13). A procedure requires that these gaps be sealed with a composition sealant on final installation of the boost protective cover; however, some gaps were not sealed. The crew reported that during the heavy rain just prior to launch they saw water on the exterior window surfaces and also observed water flowing over the windows at tower jettison. The water rivulets acted as collection sites for the exhaust residue during escape motor firing. After the water evaporated, the residue deposits remained on the surfaces of the windows.

Contamination was also noted on the inside surface of the heat-shield panes on the left-hand side and hatch windows. The contamination, which disappeared on the left-hand side window after the first day, probably resulted from water entrapped between the heat shield and pressure structure in the general area of this window. The contamination on the inside surface of the hatch heat shield window remained throughout the flight and varied in size with the thermal cycles of the spacecraft. This contamination could have resulted from either entrapped moisture in the hatch area between the heat shield and the pressure structure or from outgassing of sealant materials in this area (fig. 14-14). Such outgassing has been minimal in the past three flights because the curing processes were changed to alleviate this problem. However, a chemical analysis of the contamination on the inside surface of the hatch window has shown the concentration of silicone oils to be higher than expected. These oils are the outgassed products from the material used to seal the thermal blankets near the window.

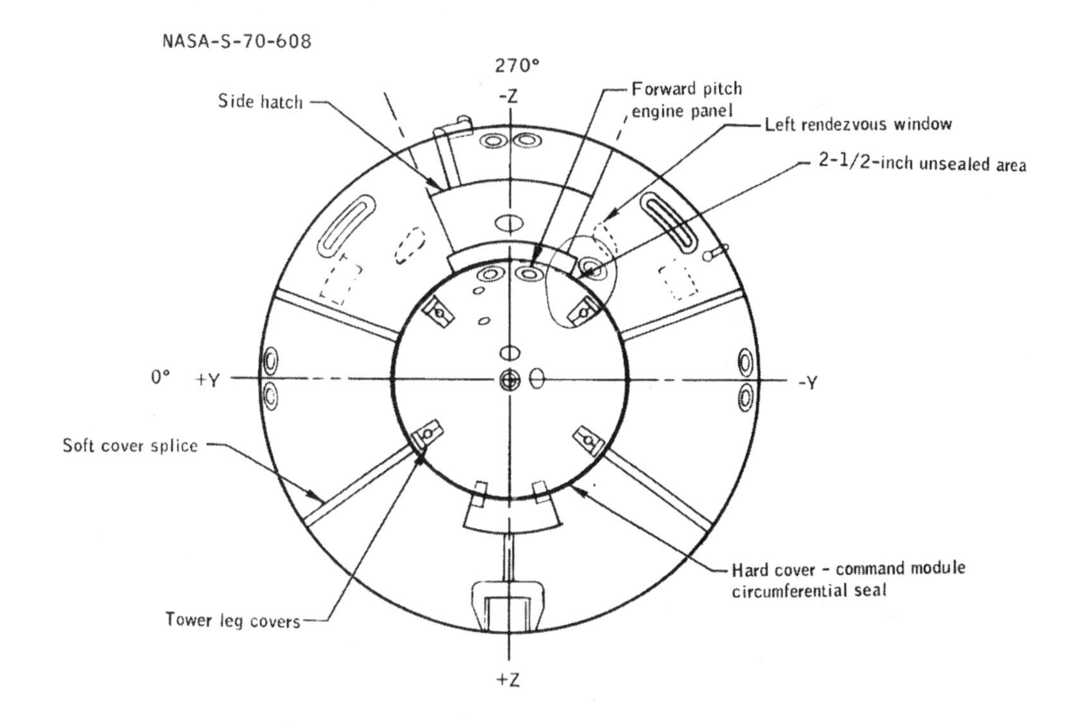


Figure 14-13 Boost protective cover view looking aft.

For Apollo 13 and subsequent spacecraft, seals will be added to the boost protective cover to prevent leakage of rain water. Prior to flight, the hatch window cavity will be purged with a 35/65-percent mixture of dry nitrogen and oxygen to remove entrapped moisture. To further alleviate the outgassing of silicone oils, the insulation material will be removed from between the outer and inner hatch windows on future spacecraft.

This anomaly is closed.

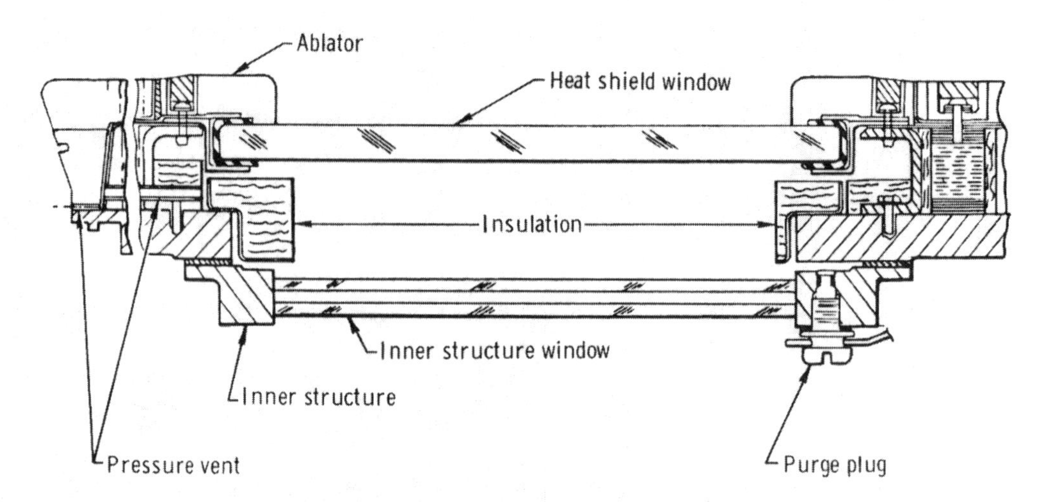


Figure 14-14 Cross section of hatch window.

### 14.1.12 Improper Deployment of VHF Recovery Antenna

During the command module descent on the main parachutes, ground plane radials 1 and 3 of VHF recovery antenna 2 (fig. 14-15) did not properly deploy. However, voice communications with the recovery forces while using this antenna were not significantly affected. Postflight examination of the antenna revealed that the cloth flap which normally covers the radials to prevent entanglement with the parachutes could be made to stick to the gusset by an adhesive substance which was inadvertently present on both the flap and the gusset. The radials would not deploy when the flap had stuck to the gusset; however, radial 1 would not always deploy, even when the flap was not stuck. A slight binding at the spring end or at the retaining clip has been experienced on radial 1.

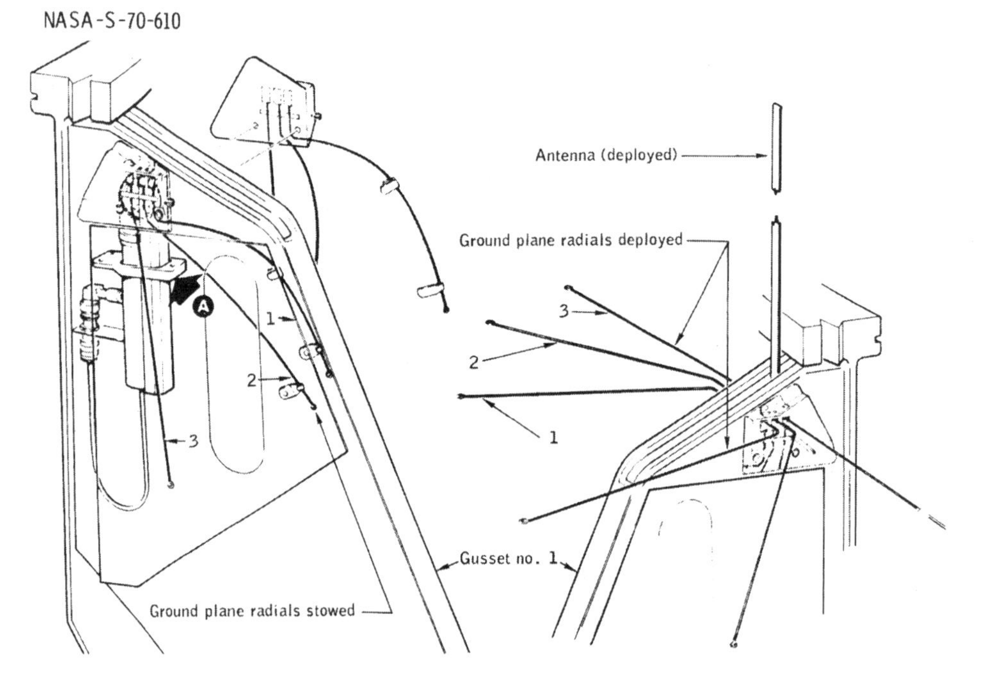


Figure 14-15 VHF recovery antenna configuration.

For Apollo 13 and subsequent missions, recovery antenna 2 will be used for recovery beacon transmissions instead of voice. However, even with no radials deployed, antenna 2 will provide a satisfactory beacon signal, with performance parameters as listed in table 14-I. Installation instructions are being studied to assure proper deployment of the radials on future flights and to insure proper removal of adhesrives.

This anomaly is closed.

**TABLE 14-1.- VHF RECOVERY CHARACTERISTICS**

| | With radials | | | Without radials | | |
|---|---|---|---|---|---|---|
| | Range, miles | Coverage [a] percent | Worst-case circuit margins | Coverage, Percent | Gain, dB | Worst-case circuit margins |
| Primary post-landing | 195 | 100 | plus 7.7 | 99 | -18 | plus 1.7 |
| Secondary post-landing | 100 | 100 | minus 3.3 | 99<br>91.5 | -18<br>-13 | minus 9.3<br>minus 4.3 |
| Primary descent | 270 | 99 | plus 4.9 | 98 | -17 | minus 0.1 |

[a]For -12 dB gain or better.

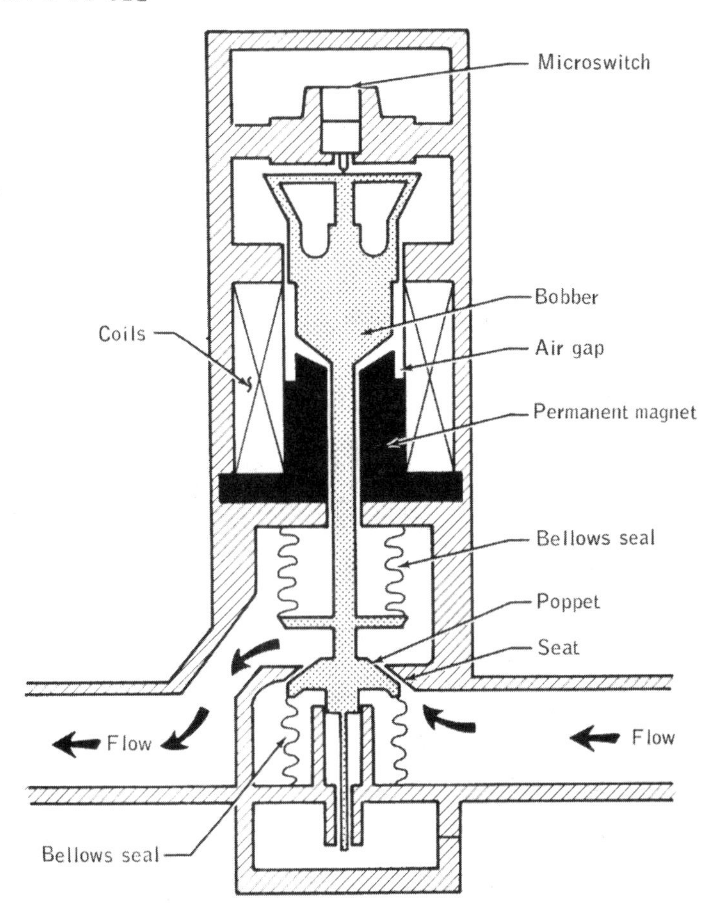


Figure 14-16 Cross sectional view of reaction control system isolation valve.

### 14.1.13 Command Module Reaction Control Isolation Valve Failure

During the postflight decontamination of the command module reaction control system, the system I oxidizer isolation valve would not remain in the closed position; however, the valve responded normally to open and close commands. This failure to remain in the closed position has been experienced when the valve bellows are distorted or damaged. The bellows hold the valve poppet in the closed position against the pull of a permanent magnet, which is used to hold the valve poppet in the open position (fig. 14-16). A damaged bellows cannot exert enough force to hold the poppet closed. Note that the valve can be held closed by applying power to the closing electromagnetic coil.

Deformed bellows are most frequently encountered when the command module reaction control system is pressurized with the isolation valves in the closed position. In this configuration, the "water hammer" effect of the fluid can deform the bellows, as was experienced in Apollo 7. However, the crew verified that the valves were opened before pressurization.

When the oxidizer isolation valve was disassembled after flight, the inlet-side bellows had been deformed enough to prevent the valve from, staying in the closed position. The bellows in the system I propellant isolation valve had also been deformed, but not enough to prevent the valve from staying closed. A review of the test and checkout history, as well as inspection records, for the Apollo 12 isolation valves indicates the valves were not degraded prior to flight. The necessity for having the valves open prior to system activation and purging will be emphasized to future crews.

This anomaly is closed.

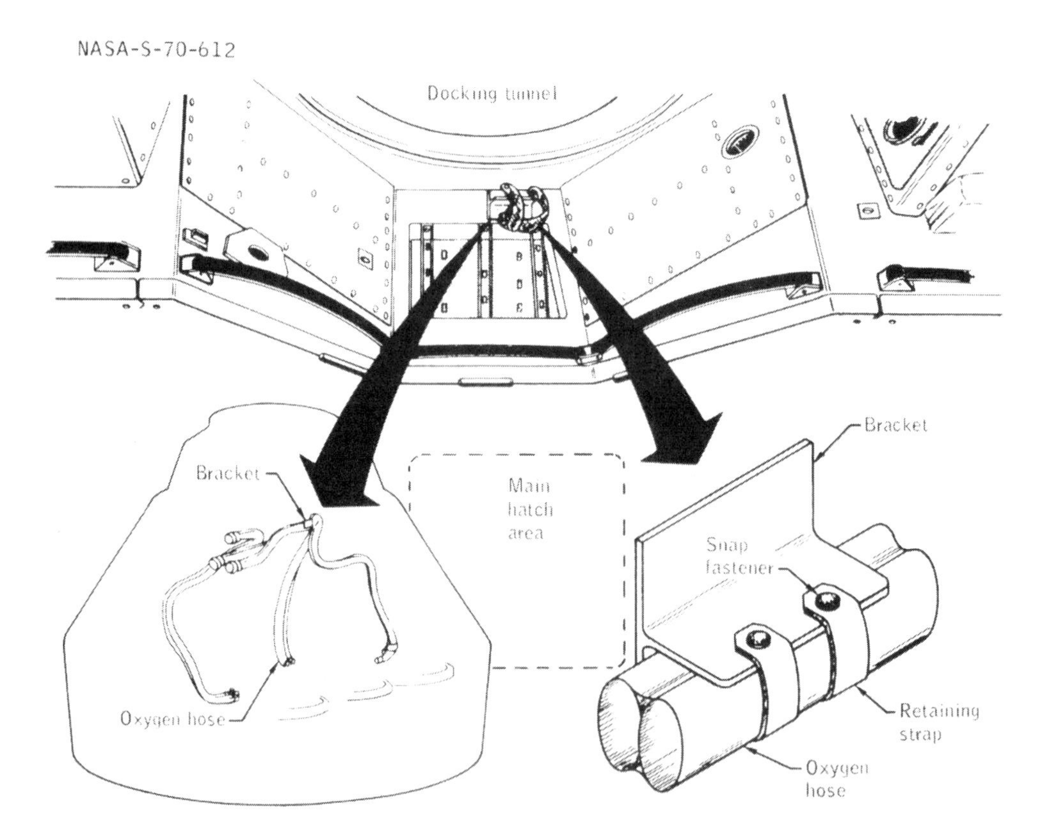


Figure 14-17 Oxygen hose retention bracket.

### 14.1.14 Oxygen Hose Retention Bracket Failure

At earth landing, an aluminum retention bracket for the oxygen hoses pulled loose from the main display panel (fig. 14-17). The bracket is bonded to the panel and supports four oxygen hoses, which are attached to the bracket by Beta cloth straps that snap to the panel.

Postflight inspection of the bracket revealed an inadequate adhesion area between the bracket and the panel. The adhesive material was not uniformly spread under the bracket, thereby creating large voids. A nonuniform application of pressure during the cure cycle is the most probable cause of this condition. Manufacturing requirements have been changed to include torque testing of the bracket to assure that a proper bond has been achieved.

This anomaly is closed.

### 14.1.15 Food Preparation Unit Water Leakage

After actuation of the hot water dispenser on the food preparation unit, the metered water flow failed to shut off completely and a slight leakage continued for 10 or 15 minutes after handle release. This leakage formed a water bubble at the end of the valve stem assembly and required blotting by the crew.

Postflight tests showed no leakage when room temperature water was dispensed through the hot water valve; however, with the heaters activated and the water temperature at the normal value of approximately 150° F, a slight leakage appeared after valve actuation. Similar results were obtained during bench tests of the unit at the vendor. Subsequent disassembly of the dispenser revealed damage in two valve 0-rings, apparently as a result of the considerable particle contamination found in the hot water valve. Most of the contamination was identified as material related to component fabrication and valve assembly and probably remained in the valve because of incomplete cleaning procedures. Since the particles were found only in the hot water valve, the contamination apparently originated entirely within that assembly and was not supplied from other parts of the water system.

Since no flight anomalies of this nature have occurred in previous spacecraft, this failure is considered to be an isolated problem and has no impact on future spacecraft.

This anomaly is closed.

### 14.1.16 Severed Lanyard on Forward Heat Shield Electrical Leads

During postflight inspection of the upper deck, the lanyard which retains the forward heat shield electrical cable had been severed, and only 18 inches of the approximately 45-inch lanyard remained. The lanyard is fabricated from natural Nomex cord with a breaking strength of approximately 600 pounds. The function of the lanyard is to provide for orderly deployment of the electrical wire bundle which connects the forward heat shield mortar cartridges and the electrical connectors on the upper deck. As the heat shield separates from the command module, the lanyard, which is anchored to the spacecraft at one end, sequentially breaks each of a series of 16- and 50-pound retainers which secure the wire bundle to the inner wall of the forward heat shield (fig. 14-18). The crew reported that parachute deployment was normal, and this is confirmed by onboard camera coverage.

Figure 14-18 Forward heat shield mortar umbilical.

Examination and comparative laboratory tests on a similar type cord disclosed that the failure is nearly identical to those which occur in lanyard knots when loaded in tension. A small flake of yellow material was found embedded in the weave of the severed end of the lanyard. Comparison of the flake with yellow Mylar tape, which is used to wrap the steel drogue riser, showed a definite similarity. Foreign material removed from the lanyard and a piece of tape from a drogue riser contained significant amount of a grayish-black material (fig. 14-19), which is believed to be deposits of a dry-film lubricant used on the steel risers.

NASA-S-70-614

Figure 14-19 Deposit on end of heat shield lanyard.

When the failed lanyard was draped over the top of the right-hand drogue mortar tube, the severed end matched the point at which the steel cable exits the mortar tube (fig. 14-20). It is therefore believed that after the lanyard broke the last retainers but prior to drogue mortar fire, the lanyard moved down over the mortar tube outboard of the drogue riser. Furthermore, when the drogue mortar was fired 1.6 seconds after heat shield jettison, the lanyard was caught over the steel cable riser and placed in sufficient tension to cause failure when the drogue was deployed. However, lanyard entanglement within the steel drogue riser would have no adverse effect on drogue function. No modification is necessary, since the lanyard satisfies its intended function prior to drogue deployment.

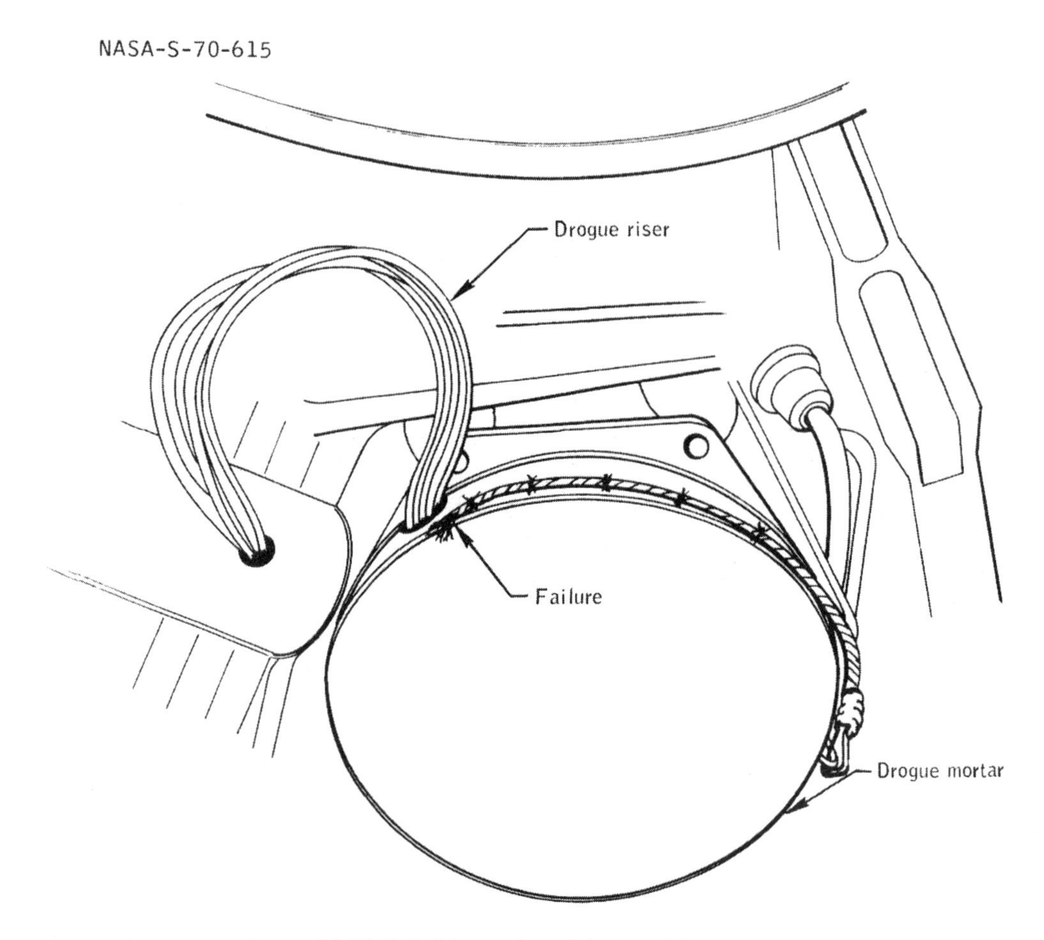


Figure 14-20 Failed lanyard at right-hand drogue mortar.

This anomaly is closed.

### 14.1.17 Instrumentation Discrepancies

*Shift in quad D helium manifold pressure.*- The measurement for reaction control quad D helium pressure indicated erroneous values throughout the flight. During the first 70 hours, the pressure exhibited a slow drift of about 14 psia upward. At approximately 160 hours, the measurement then shifted from 192 to 150 psia, followed by a second slow drift upward (fig. 14-21). Both the slow drifts upward and the jump shown on the figure tend to support the conclusion that the strain-gage bonding had weakened. The measurement is primarily used during preflight testing to indicate the helium manifold pressure downstream of parallel redundant pressure regulators and is not necessary for flight.

This anomaly is closed.

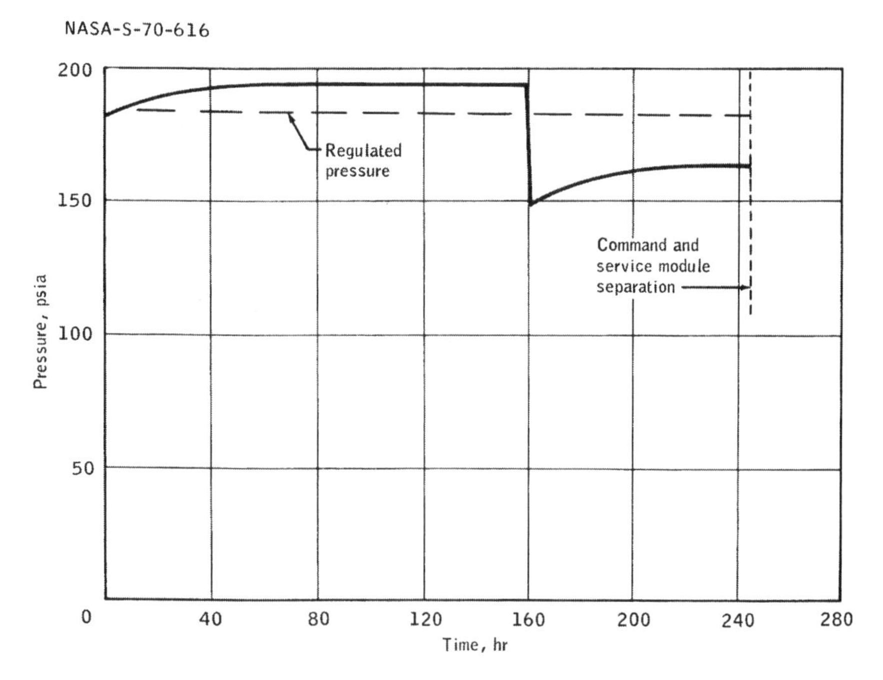


Figure 14-21 Quad D helium manifold pressure.

*Low readings from suit pressure transducer.-* The suit pressure transducer indicated low throughout the mission. The suit pressure transducer operated properly throughout the prelaunch and launch activities. When the helmets and gloves were removed after launch, the transducer indicated 0.2-psid less than cabin pressure and at approximately 22 hours the differential was 0.4 psid. A 0.4- to 0.6-psid disparity existed between the indicated suit loop and cabin pressures until the final hours of the mission (fig. 14-22).

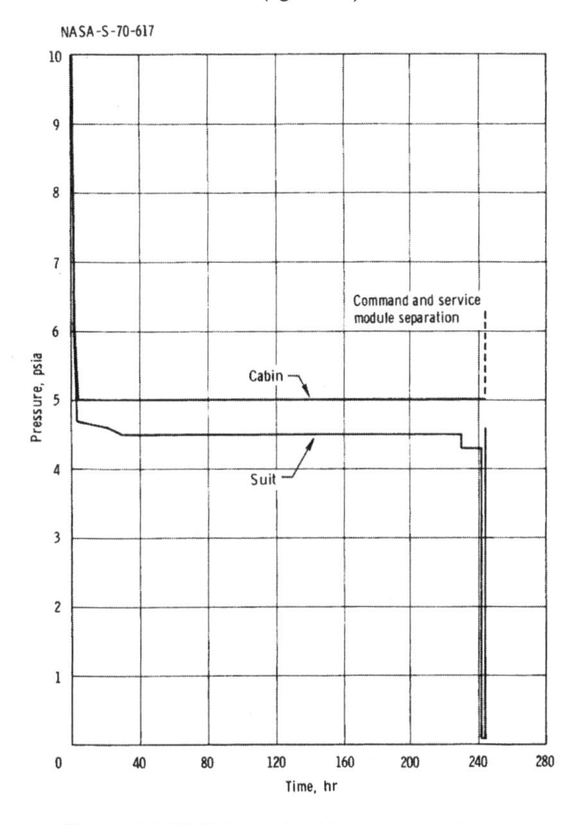


Figure 14-22 Suit and cabin pressure history.

At 241:41, the suit pressure transducer reading dropped to 0.1 psia, while cabin pressure was stable at 5.0 psia. About 3 hours later, at command module/service module separation, the transducer recovered to 4.6 psia. The transducer indicated a 0.4- to 0.5-psid discrepancy throughout entry. Postflight tests of the installed transducer repeated the flight anomaly. However, during subsequent tests of the removed transducer, the unit operated normally. The transducer was then returned to the manufacturer's facility, where flushing and disassembly revealed internal contamination from metallic nickel-plating particles. These particles could have caused an irregular transducer output by physically interfering with the Bourdon tube movement or by changing the inductance field of the unit. After the transducer was cleaned and reassembled, testing produced satisfactory operation. The noted contamination apparently resulted from either improper cleaning procedures or from self-generated particles within the unit.

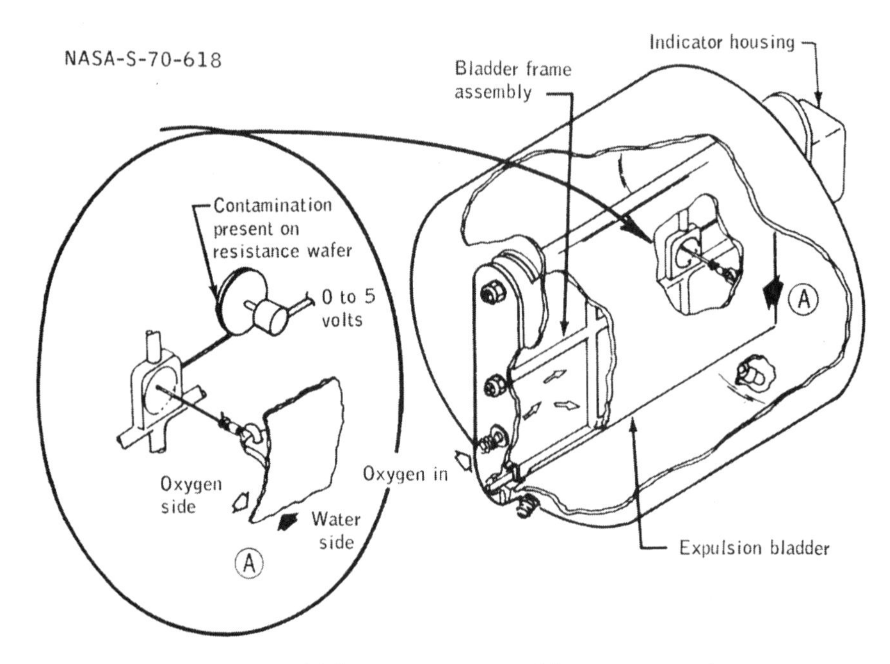


Figure 14-23 Area of failure in erratic potable water transducer.

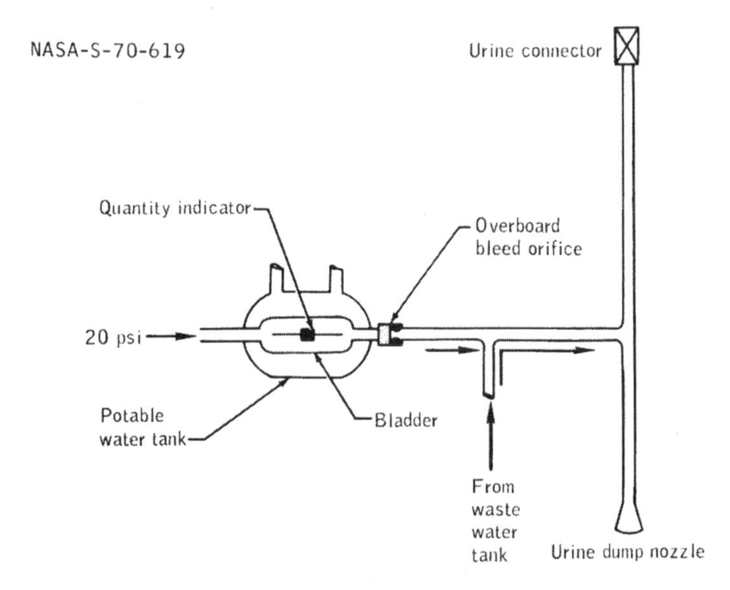


Figure 14-24 Schematic of oxygen bleed flow and overboard urine dump line.

Since previous spacecraft using both this and similar cabin pressure transducers have exhibited no problems of this type, the failure is considered to be an isolated occurrence for Apollo 12. Therefore, no impact on future spacecraft is evident.

This anomaly is closed.

*Erratic potable water quantity.*- Potable water quantity data were erratic prior to launch and also occasionally during flight. Operation of this sensor was not necessary because the known onboard water quantities were within launch specifications. Therefore, replacement, which would have required rescheduling the launch, was not performed. The sensor continued to operate erratically until about 20 hours, when the potable water tank was completely filled. The tank remained essentially full for the remainder of the flight and quantity data appeared normal during most of the mission.

Tank calibration data after flight compared favorably with those from preinstallation calibrations. Disassembly and inspection revealed that corrosion had partially obstructed the oxygen overboard bleed orifice (fig. 14-23). No evidence was found of moisture or urine contamination on components of the water measuring system.

Tests of the potentiometer reproduced the output fluctuations for wiper positions equal to approximately zero quantity (zero volts) and full quantity (5 volts). The potentiometer was disassembled and appeared clean and free of contamination except for a slight stain on the end surfaces of the resistance wafer (fig. 14-24) corresponding to wiper positions for the 0 and 5 volts. The film was removed with a water-moistened swab, but the quantity of contaminate was too small to be identified.

After removing the film, the potentiometer was reassembled and no further fluctuations were noted. Although the source of the film is unknown, acceptable alternate methods exist for determining onboard water quantities.

This anomaly is closed.

*Fuel cell 3 regulated hydrogen pressure decay.*- The fuel cell 3 regulated hydrogen pressure gradually decayed from 61.5 psia to about 59.5 psia, but remained within specification limits. The hydrogen regulator was eliminated as a possible cause of the decay, because the only regulator failure mechanism that would allow a 2-psi decay would be vent valve leakage at a rate of 2.6 pounds/hr. A 2.6-pound/hour flow rate is 38 times greater than normal for a 25-ampere individual fuel cell load and would have been easily observed on the fuel cell flowmeter.

The apparent pressure drop has been attributed to a pressure transducer failure, with the most probable failure mode being a small leak through or around the stainless steel diaphragm in the transducer (fig. 14-25). Such a leak would allow hydrogen to enter the vacuum reference chamber of the transducer, thus destroying the normal pressure differential across the diaphragm. This reduction would result in the indicated pressure decay observed during the flight. A similar transducer failure occurred during a production fuel cell pre-test checkout.

This anomaly is closed.

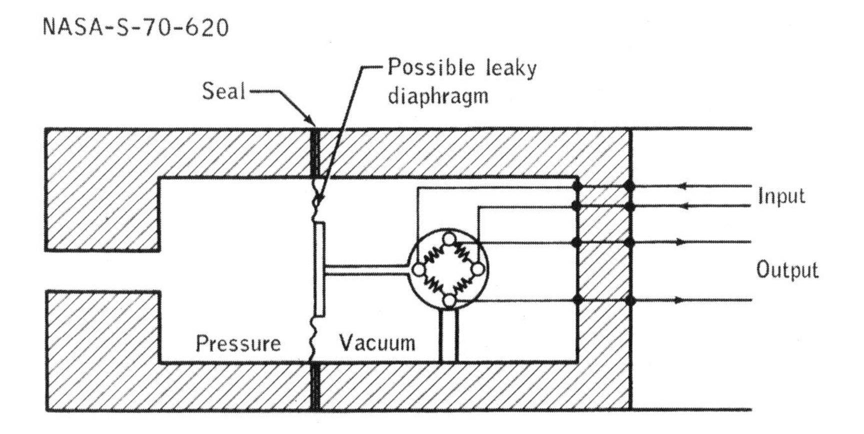


Figure 14-25 Fuel cell 3 transducer schematic.

### 14.1.18 Intermittent Tuning Fork Display

The tuning fork display on the panel 2 mission clock operated intermittently prior to and during launch. Soon after launch, the tuning fork came on and remained on throughout the remainder of the flight. This condition caused a timing error, and the mission clock had to be reset repeatedly to the correct time. The same clock had two cracks in its glass face.

Operation of the tuning fork indicates the mission clock has switched from the central-timing-equipment timing signal to an internal timing source, thus indicating loss of the central timing signal. However, the two digital event timers, which also use signals from the central timing equipment, operated correctly.

Based on previous mission clock failures, the most probable cause for this anomaly is a cracked solder joint in the cordwood construction. As seen in figure 14-26, electrical components (resistors, capacitors, diodes, etc.) are soldered between two circuit boards, and the void between the boards is filled with potting compound. The differential expansion between the potting compound and the component leads can cause solder joint cracks.

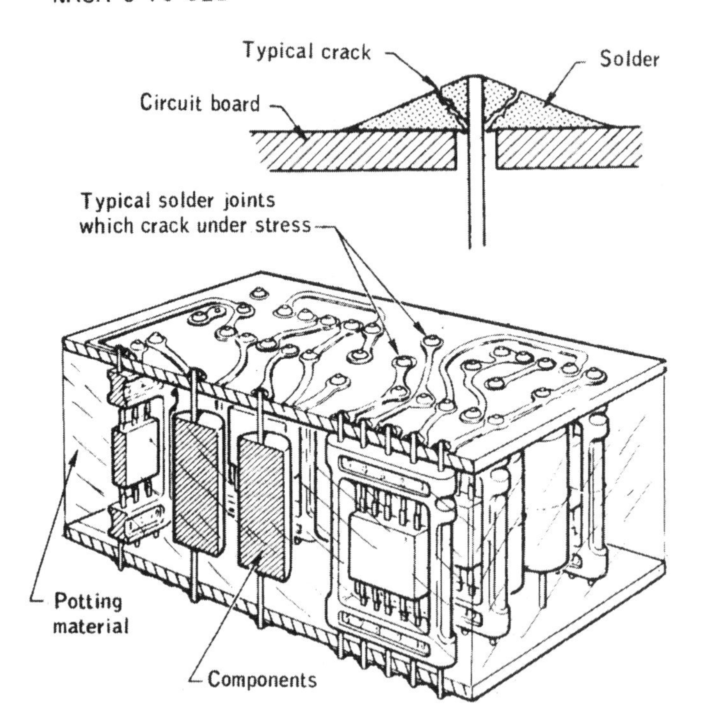


Figure 14-26 Mission timer construction.

New mission timers, which will be mechanically and electrically interchangeable with present clocks, are being developed for Apollo 13 and subsequent spacecraft. The new clock design eliminates the cordwood construction and is less susceptible to electromagnetic interference.

Both mission clocks in the Apollo 7 spacecraft and several clocks on other vehicles had cracked glass faces. The glass is bonded to the metal outer faceplate by fusing it with a ceramic frit at 1100° F. A stress induced into the glass during this process makes the glass susceptible to cracking. A clear, pressure-sensitive tape was placed over the glass face to preclude complete breakage.

This anomaly is closed.

### 14.1.19 Unacceptable VHF Communications

During ascent and rendezvous, there was a VHF communications problem between the command module and the lunar module. During this time period, there appeared to be only one problem associated with VHF voice but there were actually two separate problems. Figure 14-27 shows the VHF system as it was configured in the command module during these phases.

During ascent, there were communications from the command module to the lunar module using VHF through the lunar module aft and command module right antennas. However, beginning at 142 1/4 hours, communications from the lunar module to the command module had to be accomplished using an S-band network relay. In this case, the predicted RF signal strength (fig. 14-28) was below the sensitivity of the squelch thumbwheel setting. During the 23-minute time period following lunar module lift-off, the two vehicles had closed to a range of approximately 200 miles and the lunar module crew had switched to the forward antenna. At this point, the received signal strength at the command module improved and the Command Module Pilot began to understand the VHF voice communications.

During the time period from 142:43:00 to approximately 142:53:00, the signal strength was strong enough to maintain the squelch circuit open, as verified by flight data. During the concentric sequence initiation maneuver, the squelch was noted as dropping in and out. According to predictions, the selection of either the left or the right command module antenna did not significantly affect voice communications. During this time, the received signal strength (fig. 14-28) was approximately minus 105 dBm while using the command module right antenna and lunar module forward antenna. This figure also shows the signal strength to be minus 102 dBm or less while using the command module left antenna and lunar module forward antenna according to the flight plan. From previous tests, the squelch thumbwheel, when set at approximately 6, requires from minus 100 to 105 dBm to unsquelch the audio signal.

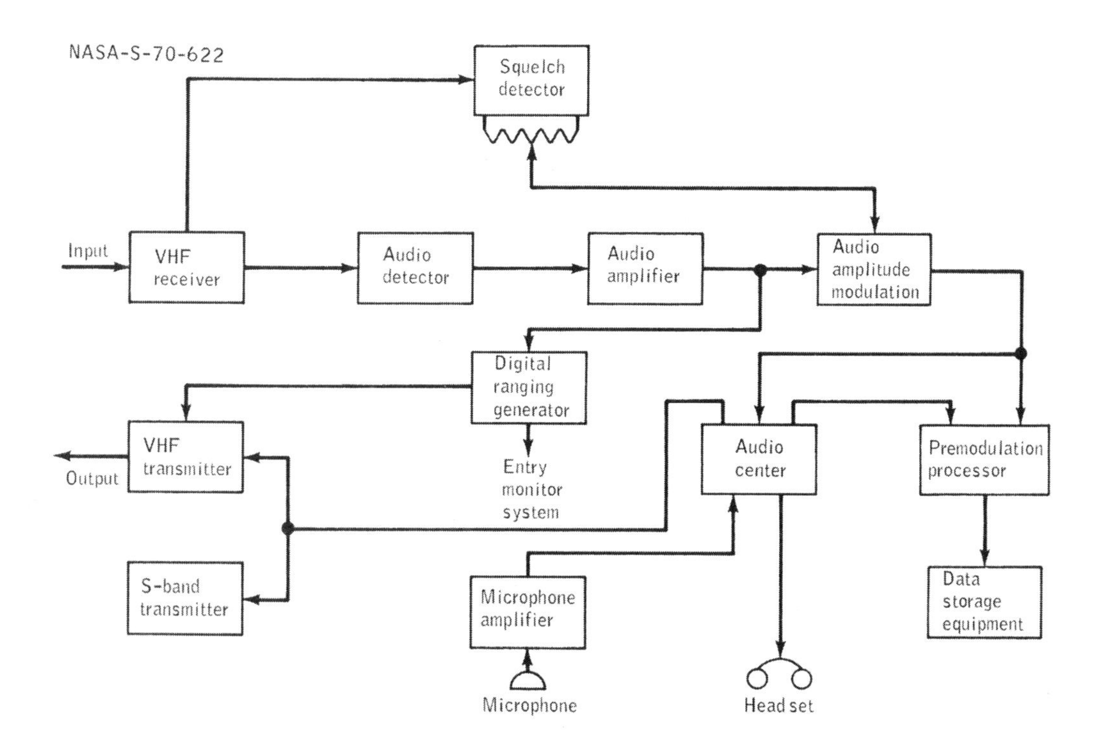


Figure 14-27 Command module VHF communications configuration.

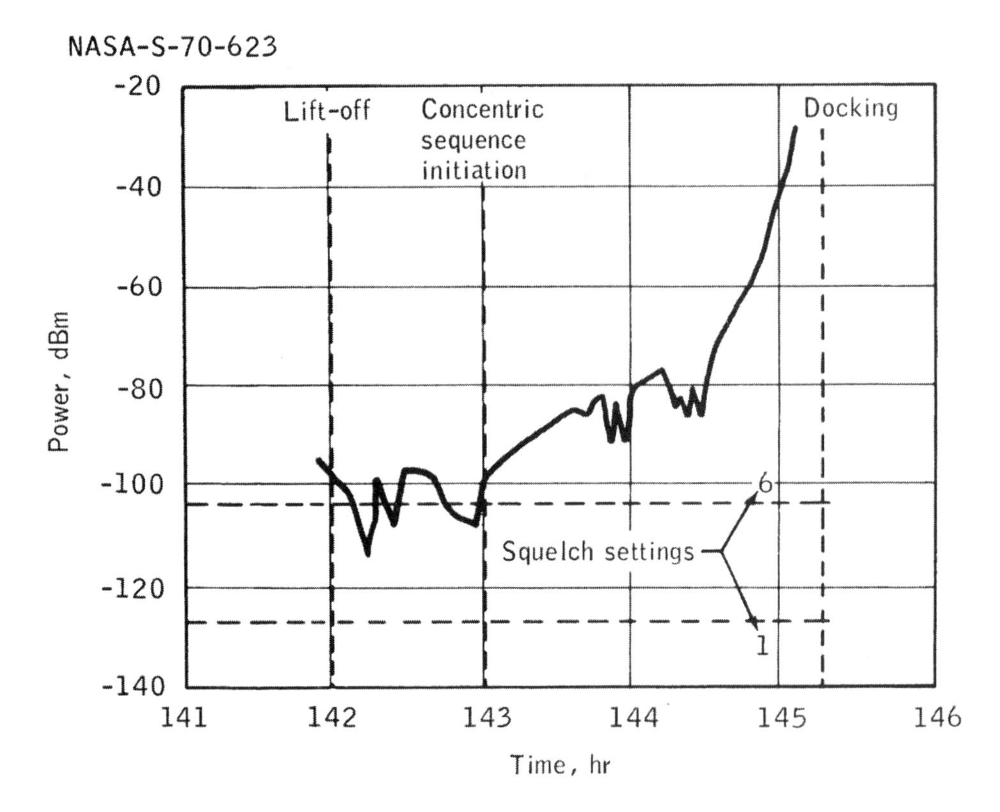


Figure 14-28 Spacecraft received power.

During the preflight checkout period, the backup crew is required to set the squelch thumbwheel to the squelch trip point and then add one increment of the thumbwheel. Since the received VHF signal is strong during this time period, there is no requirement to operate the receiver unsquelched because excessive noise would enter the system.

The VHF communications problem associated with command module reception of lunar module voice during ascent and the early part of rendezvous resulted from a low squelch-sensitivity setting in the command module VHF system. Future crews will be briefed on procedures to prevent this problem.

The second VHF voice problem during ascent and rendezvous is attributed to the use of the lightweight headset by the Command Module Pilot. S-band voice data indicate that during the time period when VHF voice to the lunar module was degraded, the voice was also degraded on the S-band link.

When the lightweight headset microphone is placed directly in front of the mouth at any distance, the headset microphone can, in effect, become a voice-cancelling circuit and reduce the voice signal level. The reduced level can then cause a voice-operated dropout of the voice operated transmitter. Such dropout did not occur at this time, because the Command Module Pilot was using the push-to-talk mode. Figure 14-29 shows the percent distortion of the lunar module received signal versue the command module audio center input, both with and without ranging. The curve shows that, in the ranging mode as the input level to the audio center decreases, distortion of the received signal increases significantly. This distortion cannot be directly related to intelligibility, but it does indicate that system performance is degraded by the low input levels.

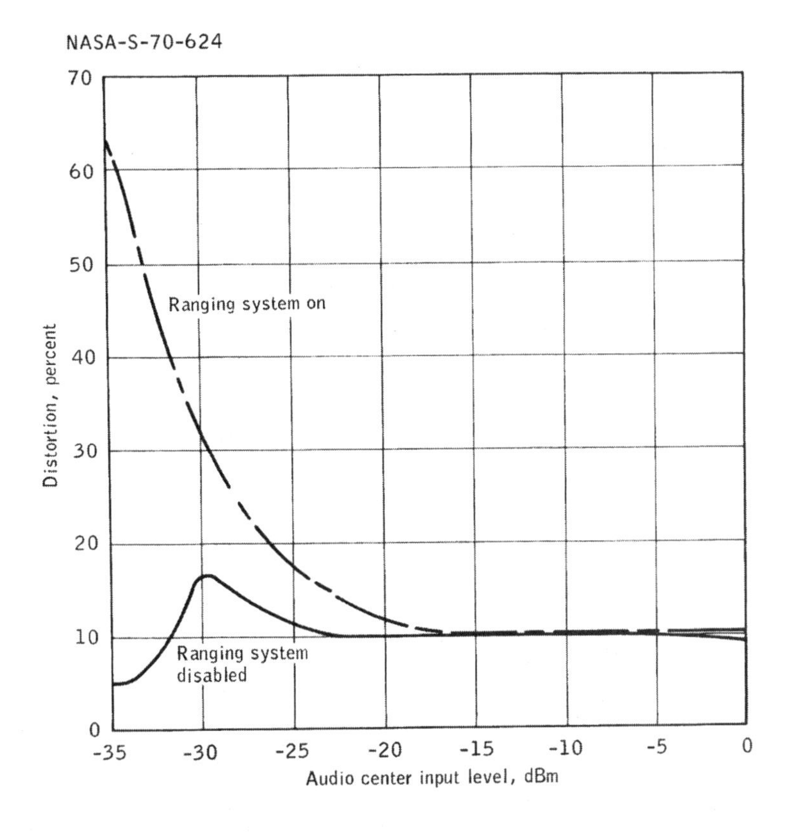


Figure 14-29 Lunar module received VHF audio distortion.

The headset microphone was designed to provide noise cancelling through mechanical spacing of the voice-capture and noise-cancelling ports (fig. 14-30). The output of the microphone amplifier is the amplified difference between the voice and noise transducer outputs. Therefore, with improper microphone placement, voice transmissions also enter the noise port, partially cancel transmissions entering the voice port, and thereby reduce the overall voice output level.

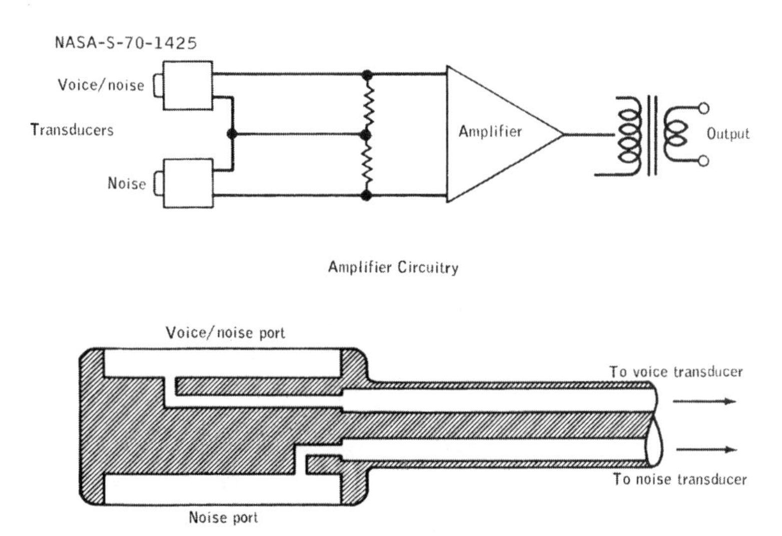


Figure 14-30 Headset microphone voice/noise parts and amplifier circuitry.

Postflight tests conducted on the headset indicate its performance to be within specification when the voice is directed properly into the voice/noise capture port, and the degraded VHF voice most probably resulted from improper placement of the lightweight headset microphone. Since there was no indication of a problem with the communications-carrier headsets, future crews will be instructed to use these headsets during critical mission phases.

This anomaly is closed.

## 14.2 LUNAR MODULE

### 14.2.1 Docking Hatch Floodlight Switch Failure

Following initial inflight checkout of the lunar module, the electrical current from the command and service module to the lunar module was approximately 1 ampere higher than expected. When the floodlight circuit breaker was turned off the current returned to the expected level.

The floodlight is controlled by a switch that is actuated by opening and closing the docking hatch in a manner similar to that for a refrigerator door. The crew checked the operation of the hatch switch and verified floodlight operation by manually depressing the plunger. However, the hatch did not depress the plunger sufficiently to actuate the switch.

The method of setting plunger travel was found to be inadequate, and a new procedure has been incorporated to specify a plunger travel of 0.120 (±-0.005) inch.

This anomaly is closed.

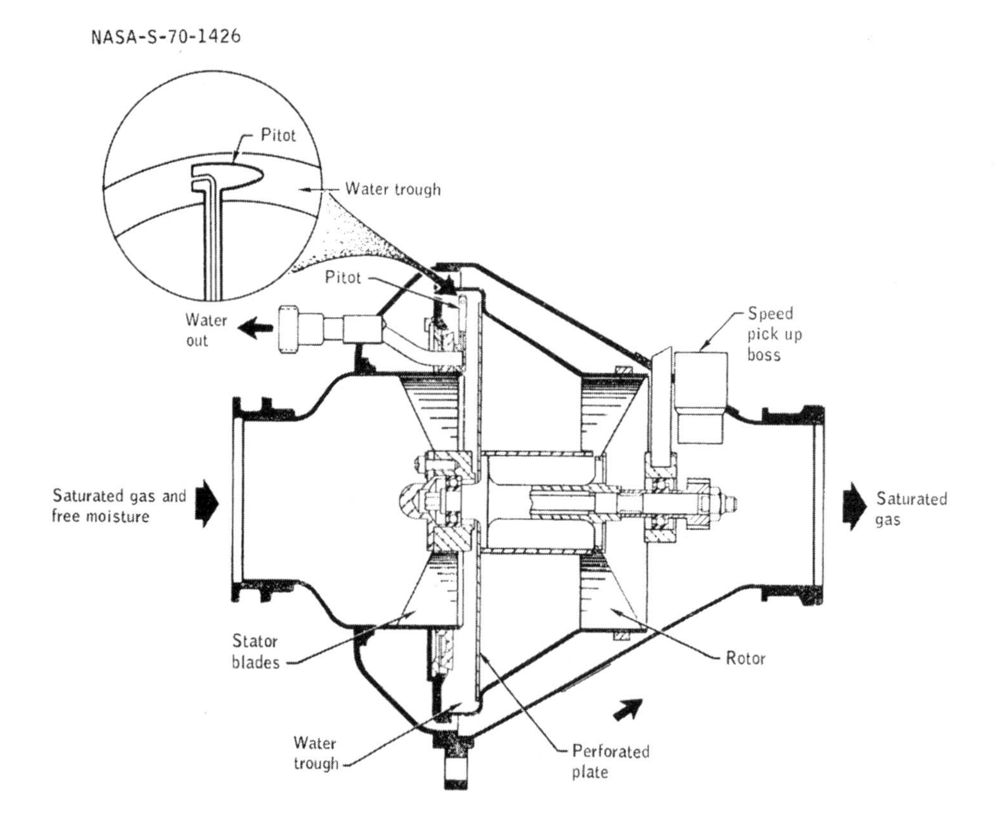


Figure 14-31 Water separator and pitot configuration.

### 14.2.2 Water in the Suit Loop

During preparations for the first extravehicular activity, water was reported coming from both suit inlet hoses when disconnected.

After the first extravehicular activity, the Commander reported that his boots had water in them and that the suit inlet hose was delivering cold moist air when disconnected. The Lunar Module Pilot also noted drops of water in his inlet hose. The water separators were switched with no improvement in the free water condition. Prior to the sleep period, the water was drying in the Commander's suit, and there was no further problem with water in the suits.

Two possibilities exist for introducing free water into the suit loop: water may have been bypassing the water separator, or water may have been condensing out of the gas in the suit hoses.

The water separator speed indication was above the upper limit (in excess of 3600 revolutions per minute) for about 50 percent of the mission. Since the water separator is a gas-driven centrifugal pump, this high speed indicates a higher than normal gas flow through the separator. Tests have shown that, at separator speeds in excess of 3700 revolutions per minute, water splashing occurs at the picot tube (fig. 14-31) allowing water to bypass the separator.

Since the coolant lines for the liquid cooling garment are adjacent to the oxygen hoses in each crewman umbilical assembly, condensation in these hoses was investigated. The analysis showed that with the flight conditions, condensation did not take place in the suit hoses.

For Apollo 13 and subsequent missions, a flow limiter (fig. 14-32) will be added to the primary lithium hydroxide canister to reduce suitloop gas flow and consequently limit the separator speed to within the no-splash range. The flow limiter provides restriction of flow equivalent to the secondary canister. If necessary, this added resistance can be removed in flight.

This anomaly is closed.

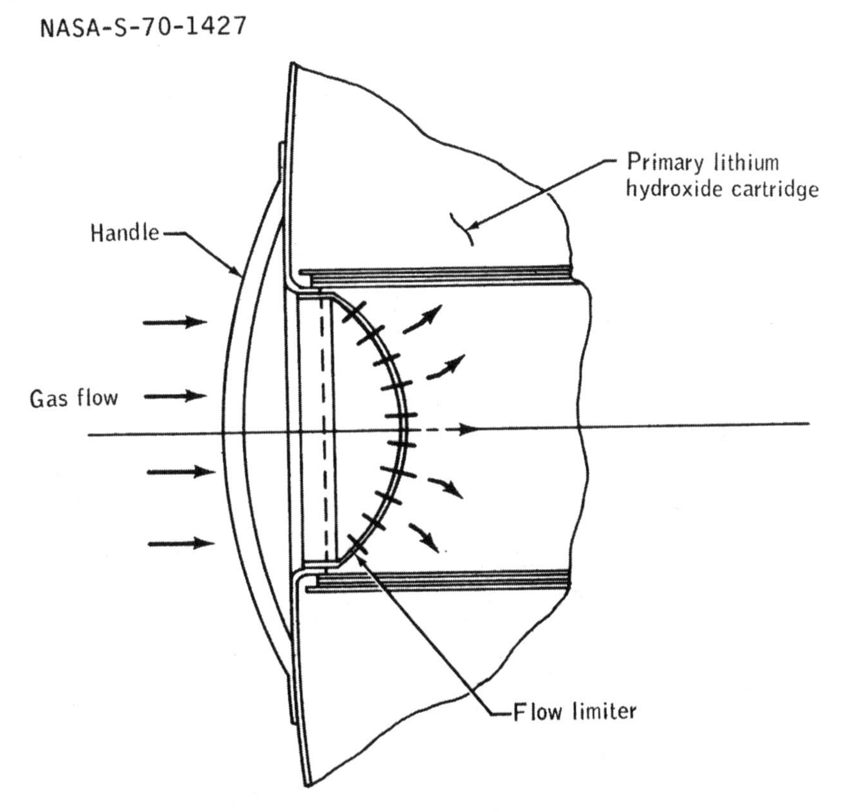


Figure 14-32 Suit circuit flow limiter.

### 14.2.3 Carbon Dioxide Sensor Malfunction

Following lunar lift-off, the crew reported a master alarm at about the time of ascent-engine shutdown. Ground data show a short-duration spike in the indicated carbon dioxide partial pressure at that time. During the second pass behind the moon following lift-off, the crew reported that the indicated carbon dioxide partial pressure again tripped the carbon dioxide high partial pressure light and master alarm. The crew selected the secondary lithium hydroxide canister at this time. The primary canister was later reselected at the request of ground controllers. The crew later reported that erratic carbon dioxide indications occurred while using either the primary or secondary lithium hydroxide canisters.

The carbon dioxide sensor is sensitive to free water, and the malfunction was probably caused either by water from the water separator sump tank entering the sensor or by water bypassing the water separator and entering the sensor. The water separator sump tank vent line joins the carbon dioxide sensor inlet sense line (fig. 14-33). This vent line has been rerouted for Apollo 13 and subsequent vehicles.

This anomaly is closed.

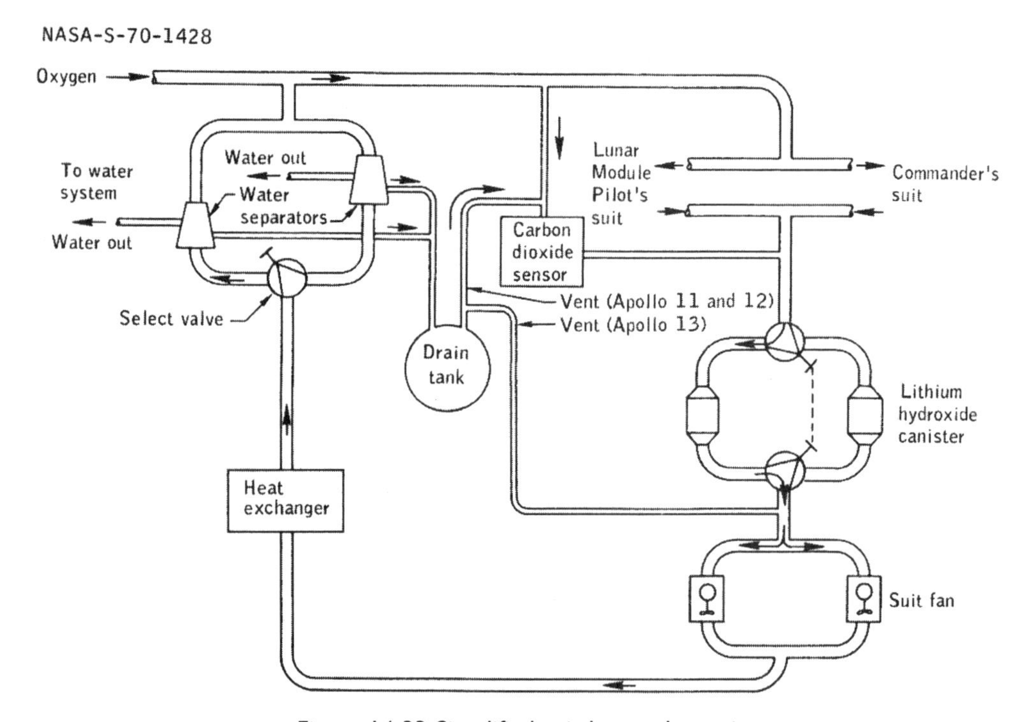


Figure 14-33 Simplified suit loop schematic.

### 14.2.4 Tracking Light Failure

At the beginning of the second darkness pass after lunar lift-off, the crew reported that the tracking light had failed. Subsequent cycling of the light switch indicated that power consumption was normal, indicating the high-voltage section of the light had experienced a corona failure.

The characteristics of the failure are very similar to failures that were experienced on Apollo 9 and in ground testing. These previous failures were attributed to corona in the high voltage section of the light. After the Apollo 9 failure, numerous design modifications were made to reduce the corona problems. Lights with these modifications successfully completed qualification testing and a lunar flight simulation and operated satisfactorily on Apollo 11.

Tests indicate that off-axis solar impingement on the flash head reflector can cause temperatures on the flash head potting as great as 500° F, which could degrade the potting compound enough to cause a corona.

For Apollo 13 and subsequent missions, the tracking light will be redesigned to reduce the 4000-volt voltage source to 2000 volts, and flash head potting will be protected from direct solar impingement. The 1-hour acceptance test operating time will be increased to 5 hours so that units with defective potting can be identified.

This anomaly is closed.

### 14.2.5 Equipment Compartment Handle Did Not Release

During the initial egress, the modularized equipment stowage assembly was to be deployed by pulling a special D-ring handle. Although the Commander was unable to release the handle from the support bracket, it could be rotated in its bracket. The equipment compartment was subsequently deployed by pulling on the bellcrank cable, which attaches to the center of the D-ring handle. A retention pin at the bottom of the D-ring handle plugs into a socket in the retaining bracket (fig. 14-34). This socket contains a ball detent mechanism which holds the D-ring to the bracket. Apparently, either there was binding in the ball detent or the crewman pulled on the D-ring handle at such an angle that a lateral load was applied to the retention pin, causing it to bind in the retention socket.

For Apollo 13 and subsequent, the D-ring will be deleted and a loop will be clamped to the end of the deployment cable. The loop will be retained using the same type of pin presently installed to retain the safety wire (fig. 14-34).

This anomaly is closed.

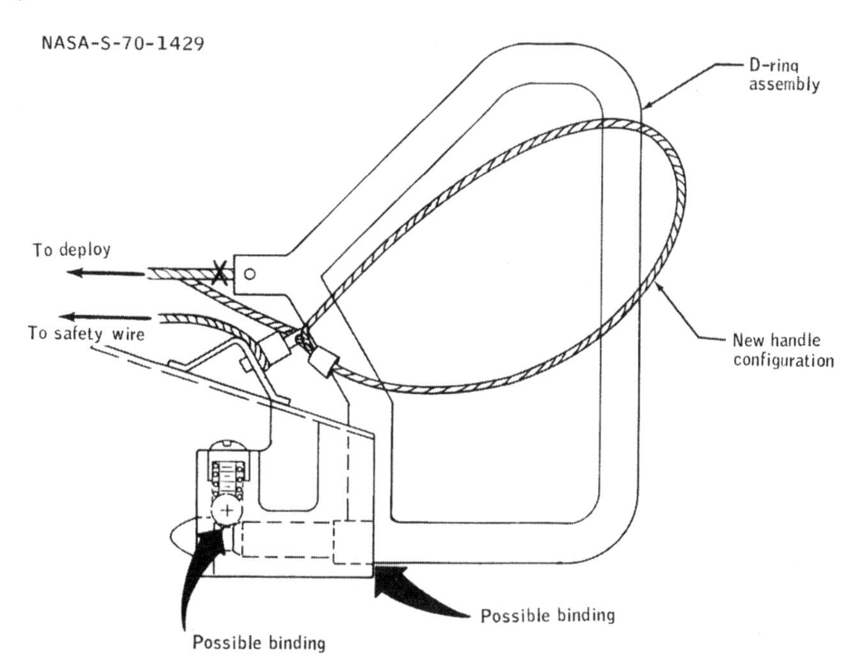


Figure 14-34 Deployment handle (D-ring) on the modular equipment storage assembly.

### 14.2.6 Torn Forward Hatch Thermal Shield

During egress, the Commander's portable life support system came in contact with and tore the hatch micrometeoroid shield (fig. 14-35). Such a tear could represent a potential hazard to the suit. For Apollo 13 and subsequent, the thermal shield thickness will generally be increased from 0.004 to 0.010 inch. At the standoff, however, the shield thickness will be increased from 0.020 to 0.040 inch. In addition, the diameter of the shield mounting holes will be increased from 0.375 to 0.5 inch (fig. 14-36). These modifications should strengthen the shield sufficiently to prevent tearing in any future contacts by the egressing crewmen.

This anomaly is closed.

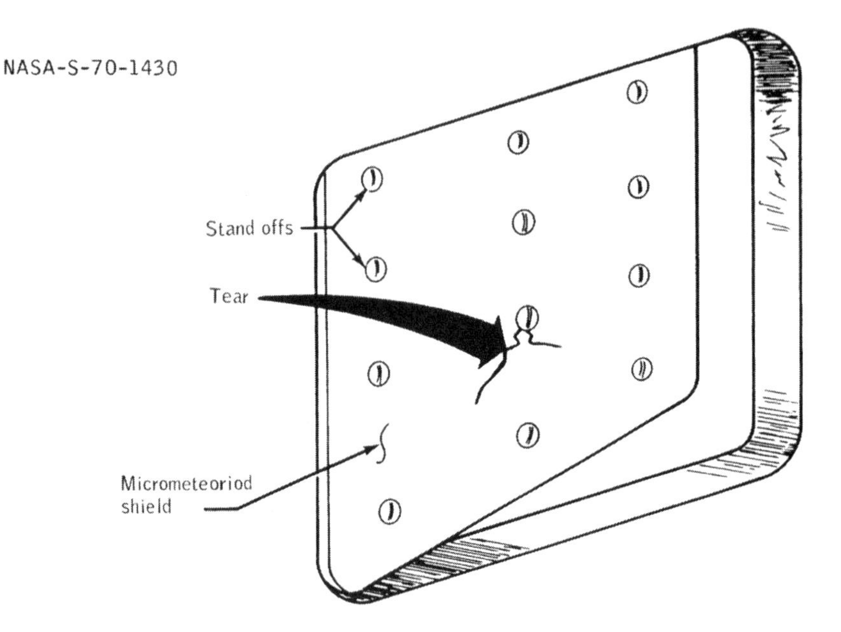


Figure 14-35 Tear in forward outer skin.

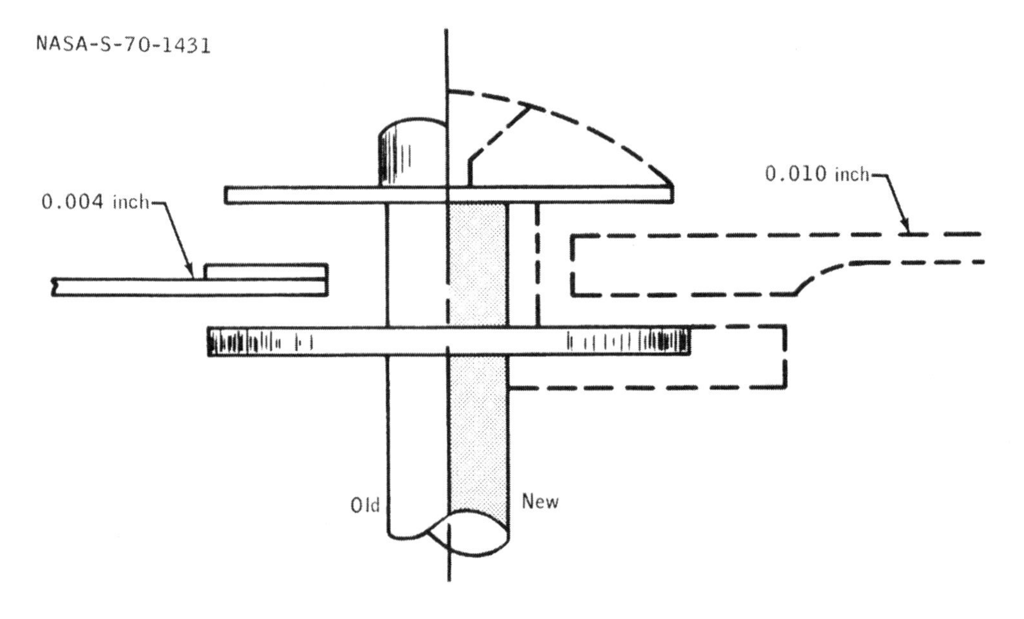


Figure 14-36 Forward hatch standoff configuration.

### 14.2.7 Early Illumination of the Low-Level Descent Light

The low-level light for descent propulsion propellant quantities illuminated about 25 seconds early. The low-level light is activated and remains latched on when any one of the four low-level point sensors (one in each propellant tank) is uncovered (fig. 14-37).

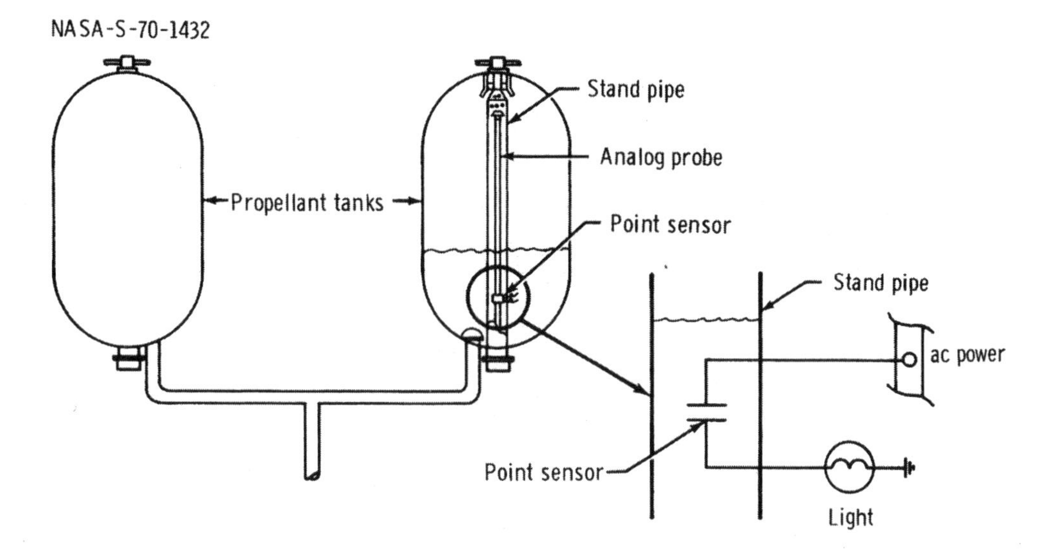


Figure 14-37 Descent propellant tank low-level sensor schematic.

At low-level light activation, the gaging system indicated that fuel tank 2 had a mean propellant quantity of 6.7 percent. In addition, it had about a 2.3-percent peak-to-peak oscillation (fig. 14-38), probably caused by propellant slosh, which continued for some time after landing. The other three tank readings experienced similar oscillations, although at a slightly higher mean quantity level. One of the four low-level point sensors, probably fuel tank 2, uncovered momentarily because of propellant slosh, causing the low-level light to latch on.

The quantity warning light should illuminate when the lowest indicated propellant level remaining in any tank reaches a value of 5.6 ±1/4 percent. Since the light came on when the averaged quantity measurement indicated 6.7 percent with an oscillation of ±1.1 percent, the lowest excursion of the quantity reading was 5.6 percent and the display operated properly. The averaged propellant quantity reached 5.6 percent about 25 seconds later.

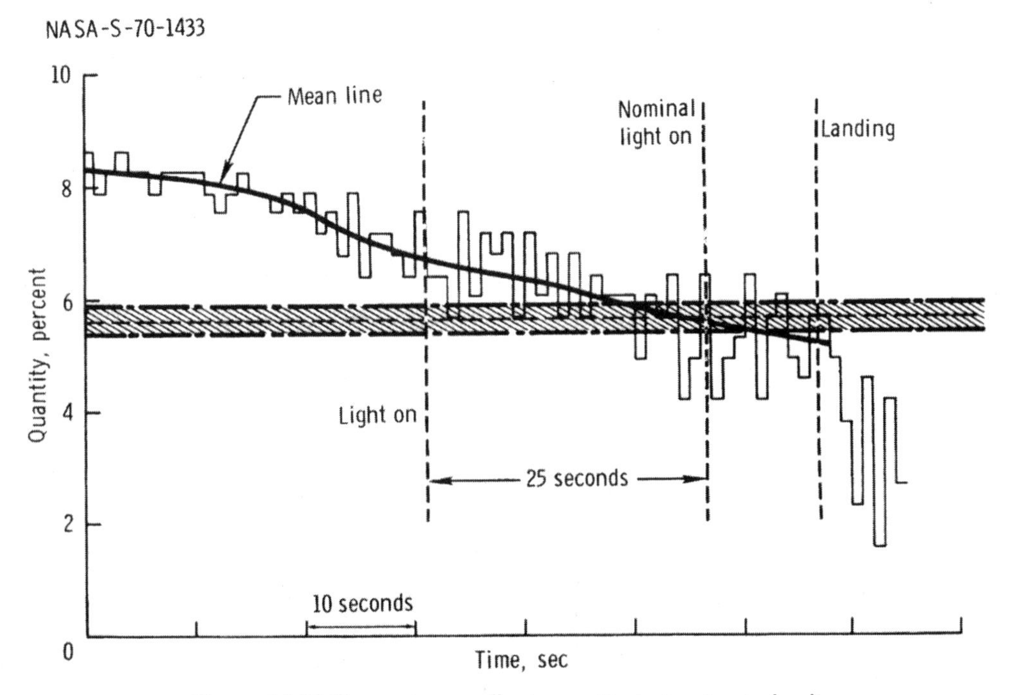


Figure 14-38 Descent propellant quantity just prior to landing.

For Apollo 13, the quantity measurements for the four descent propellant tanks have been increased in sampling rate from 1 to 100 samples per second. These data will be averaged automatically and used to determine the low-level point from which the remaining firing time can be calculated. The 100 samples per second rate will provide data that will permit an understanding of the particular dynamics of the fluid in the tanks.

This anomaly is closed.

## 14.3 GOVERNMENT FURNISHED EQUIPMENT

### 14.3.1 Color Television Failure

The color television camera provided satisfactory television coverage for approximately 40 minutes at the beginning of the first extravehicular activity. Thereafter, the video display showed only white in an irregular pattern in the upper part of the picture and black in the remainder. The camera was turned off after repeated attempts by the crew to restore a satisfactory picture.

Ground tests using an Apollo-type image sensor (secondary electron conducting vidicon tube) exposed the camera system to extreme light levels. The resulting image on a monitor was very similar to that seen after the flight camera failure.

After decontamination and cleaning, the flight camera was inspected and power was applied. The image, as viewed on a monitor, was the same as that last seen from the lunar surface. The automatic light-level control circuit was disabled by cutting one wire. The camera then reproduced good scene detail in that area of the picture which had previously been black, verifying that the black area of the target was undamaged, as shown in figure 14-39. This finding also proved that the combination of normal automatic light control action and a damaged image-tube target caused the loss of picture. In the process of moving the camera on the lunar surface, a portion of the target in the secondary-electron conductivity vidicon must have received a high solar input, either directly from the sun or from some highly reflective surface. That portion of the target was destroyed, as was evidenced by the white appearance of the upper part of the picture.

Training and operational procedures, including the use of a lens cap, are being changed to reduce the possibility of exposing the image sensor to extreme light levels. In addition, design changes are being considered to include automatic protection, such as the use of an image sensor which is less susceptible to damage from intense light levels.

This anomaly is closed.

### 14.3.2 Intermittent 16-mm Camera During Ascent

The 16-mm camera was turned on just before lift-off, but it stopped after a brief period of operation. During ascent, it was activated two additional times, and each time it stopped after 20 or 30 seconds of operation. During rendezvous, the camera was operated by constantly depressing the triggering button, thereby overriding the automatic shutoff.

The camera had performed satisfactorily for more than 8-1/2 hours during separation, descent, panoramic views of the lunar surface, and continuously throughout the two extravehicular activities. The camera is certified for 10 hours of operation in a vacuum.

Although postflight tests showed the 16-mm camera and magazine to be in satisfactory operating condition, the characteristic sensitivity of the magazine interlock microswitch installation is such that the operating limits of the switch could cause intermittent actuation. The intermittent operation was duplicated on the flight and similar equipment by the application of pressure to the end of the magazine. The problem will be resolved by changing the interlock switch (fig. 14-40) to a configuration that is much less sensitive to variation in switch settings.

This anomaly is closed.

NASA-S-70-1434

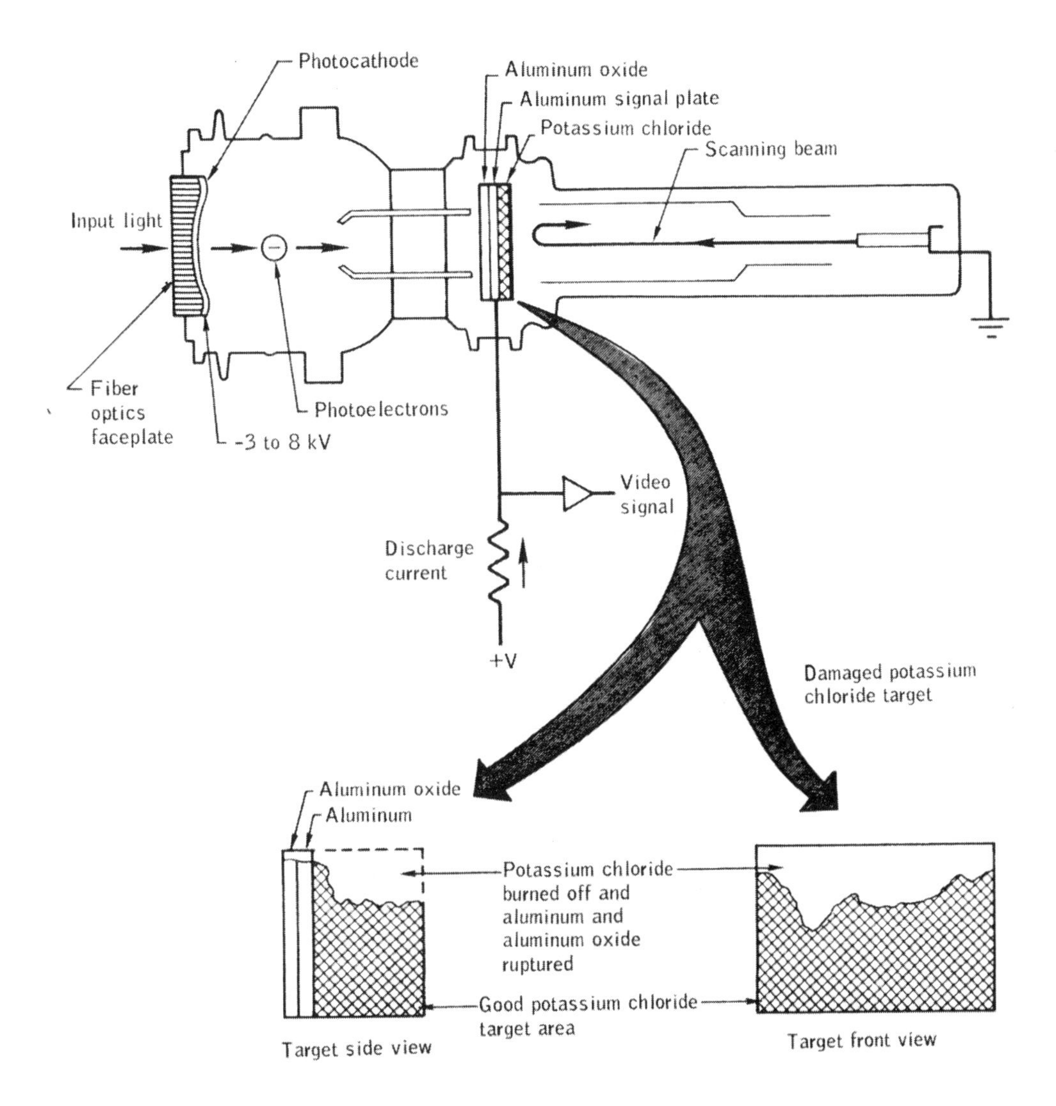


Figure 14-39 Secondary electron conductivity tube in the color television.

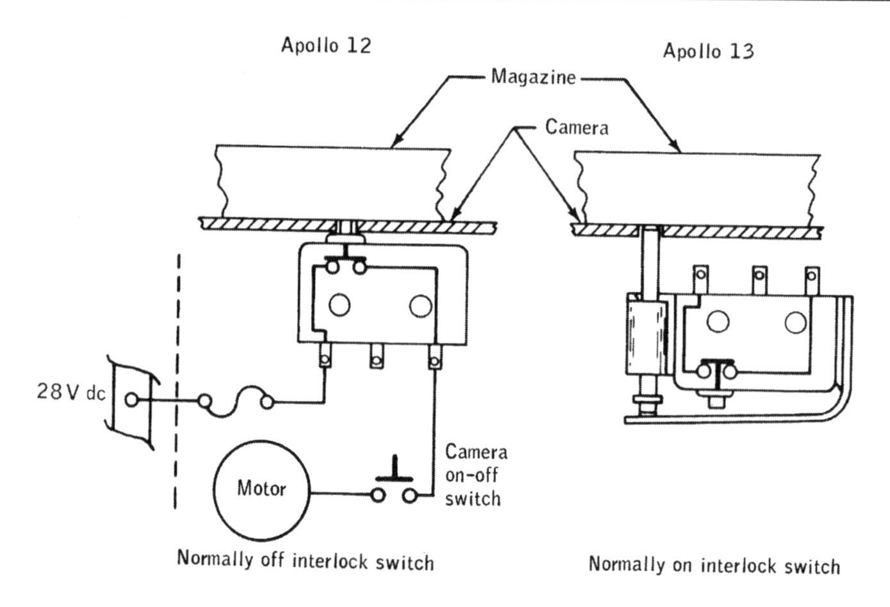


Figure 14-40 Sequence camera interlock switch modification.

### 14.3.3 Difficulty in Removing the Radioisotope Fuel Capsule

The crew experienced difficulty in removing the radioisotope fuel capsule from the fuel cask assembly during deployment of the Apollo lunar surface experiments package.

Thermal tests and analyses show that dimensional tolerances can diminish with temperature and result in binding between the latch fitting (C-ring) on the cask and the contact surface of the backplate on the fuel capsule (fig. 14-41). The longitudinal contact distance for these two surfaces is approximately 0.6 inch, and extraction was easily accomplished once this distance was negotiated.

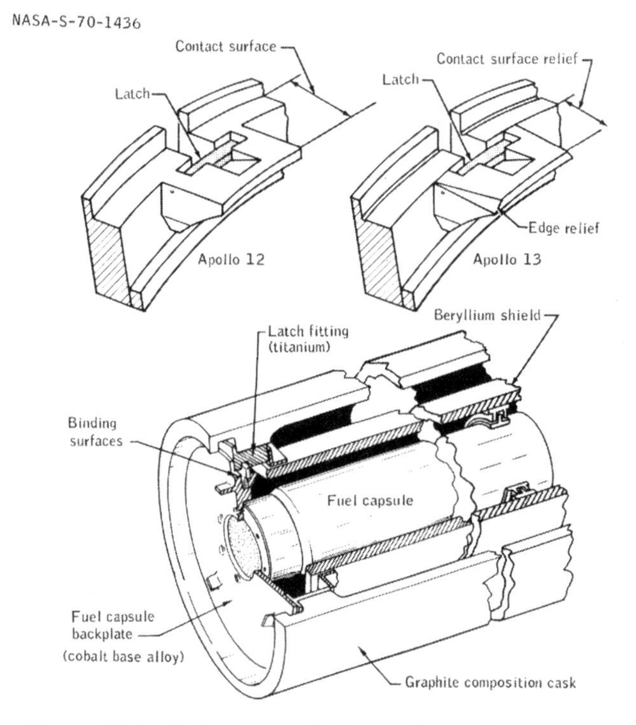


Figure 14-41 Radioisotope fuel capsule configuration.

As a result of the dimensional checks, the thermal tests, and analyses performed with both the qualification and Apollo 13 flight hardware, the contact surfaces of the fuel capsule backplates are being reworked as indicated in the figure. The outside diameter of the 0.10-inch long contact surface, while remaining within design limits, may be reduced as much as 0.005 inch for ease of capsule extraction. All existing capsule backplates will be reworked in this manner.

This anomaly is closed.

### 14.3.4 Difficulty in Deploying the Passive Seismometer

The lunar surface material at the deployment site for the passive seismic experiment was soft and irregular, and a crewman had to use his boots to tamp a depression in the surface material in preparation for deployment. This procedure, however, was in accordance with the preflight plan for this surface condition.

The thermal shroud tended to delaminate and rise up off the lunar surface. This condition had been anticipated, and lunar soil was placed on the periphery of the shroud to hold it down. When this operation proved difficult, tie-down bolts, which had been removed from the pallet during deployment of the experiments package, were placed on the shroud with satisfactory results (fig. 14-42) .

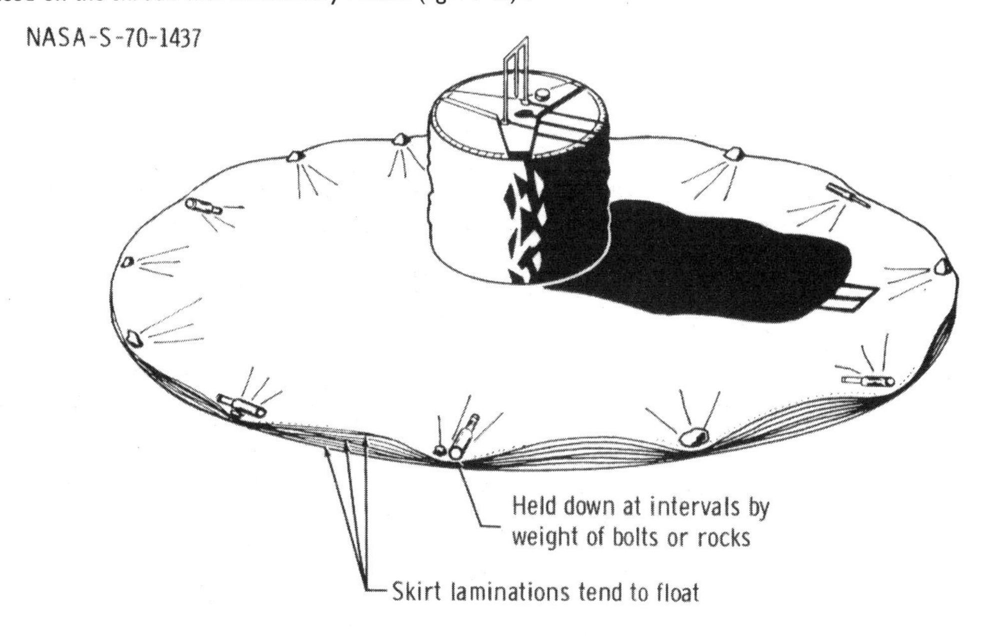


Figure 14-42 Passive seismic experiment deployed.

For Apollo 13 and subsequent spacecraft, the shroud laminations will be spot-sewed together at intervals around the periphery, a weight will be sewed to each of the six attach-pullout points on the shroud, and a 5-foot diameter Teflon blanket will be added for thermal control to decrease solar degradation.

This anomaly is closed.

### 14.3.5 Difficulty in Deploying the Cold Cathode Ion Gage

The cold cathode ion gage would not remain upright when deployed. Its final position was on its back with the sensor aperture at an angle of approximately 60 degrees from the horizontal but was satisfactory (fig. 14-43).

The cable connecting the cold cathode ion gage with the suprathermal ion detector was quite stiff. The combination of the spring effect in the cable, the reduced weight of the cold cathode ion gage under lunar gravity, and the softness of the lunar surface was apparently sufficient to cause the equipment instability during deployment. Final positioning of the equipment requires that the sensor aperture does not point directly at the surface nor directly at other experiment package components. The final positioning fulfilled this requirement.

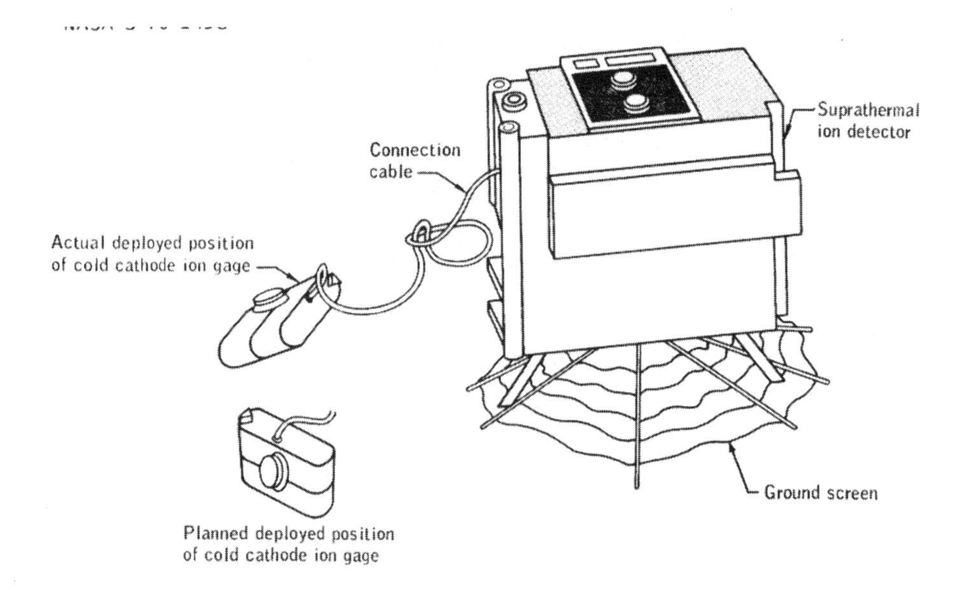


Figure 14-43 Cold cathode gage deployed.

The combination of the suprathermal ion detector with the cold cathode ion gage will not be included for Apollo 13. For Apollo 14 this equipment will be flown, and the wires of the connecting cable will be tied at 6-inch intervals instead of being wrapped with heavy Mylar tape. This modification not only reduces cable stiffness by 70 percent, which decreases the spring effect, but also decreases cable bulkiness to permit easier stowage.

This anomaly is closed.

### 14.3.6 Unsatisfactory Tool Carrier Bag Retention

At the beginning of extravehicular activity, the empty tool carrier collection bag tended to rise out of the tool carrier until some lunar surface soil was put in to hold it down. The bag is attached to the carrier structure by three aluminum spring clips (fig. 14-44). The weight of the loaded bag is shared by these clips and three hangers. The retention force is limited so that the loaded bag may be easily lifted out of the carrier.

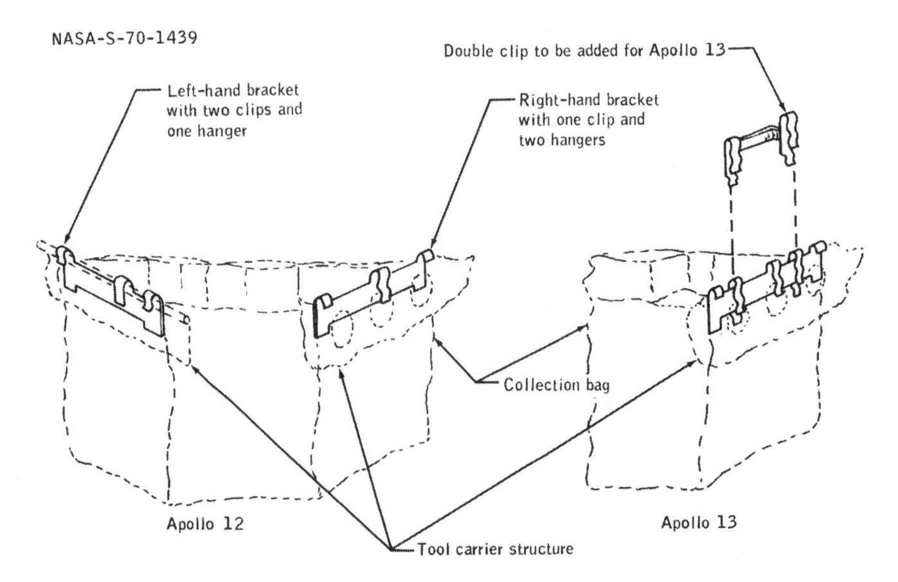


Figure 14-44 Tool carrier collection bag retention.

The retention characteristics of the left side, with two spring clips over the 0.37-inch diameter rolled bead of the carrier structure, is satisfactory. However, the single spring clip over the 0.18-inch lip of the carrier on the right side did not provide sufficient positive retention. A separate double spring clip, which reaches over both the bag hanger and tool carrier structure, will be added for Apollo 13 to provide the necessary retention force as shown in the figure.

This anomaly is closed.

### 14.3.7 Intermittent Counting on the Command Module 70-mm Camera

During landmark tracking using 70-mm camera with the 500-mm lens, the magazine opened up and the counter did not agree with the crew count. The crew had inadvertently actuated the mechanism which opens the magazine, allowing the entire film holder portion of the magazine to come out of the magazine housing. When the film holder is not inserted properly and not locked in the magazine, the film drive mechanism will become disengaged and the camera may not transport an entire frame of film each time. Overlapping exposed frames of film from this magazine indicate that this condition occurred. Since there is no requirement to remove film during the mission, tape will be placed over the retracted film release knob after loading the magazine, and proper frame counting should be preserved.

This anomaly is closed.

### 14.3.8 Suit Pressure Pulses

During the second extravehicular period, the Lunar Module Pilot indicated that he felt something which could have been two pressure pulses in the pressure garment assembly, but he could not determine whether the pulses were increases or decreases in pressure. During the first pressure pulse, the cuff gage indication for the pressure garment assembly was normal. The mission time for the reported pressure pulse, based on a sharp rise in the Lunar Module Pilot's heart rate, was determined to be between 133:09:00 and 133:12:00.

Although suit data were reviewed throughout both extravehicular periods, there was no evidence of a pressure pulse. In particular, data from 133:06:16 until 133:12:29 showed that the pressure garment assembly pressure remained constant at 3.86 psi.

A sudden pressure increase must come from the pressure regulator in the portable life support system. The increased pressure would remain high until the suit pressure returned to normal, but at a slow rate which would not exceed 0.3 psi/min. For a measurable pulse increase of 0.1 psi, this decay would take 20 seconds and would be detectable in telemetry data. A sudden pressure decrease indicates a momentary leak in the system. For a measurable decrease of 0.1 psi, the portable life support system maximum makeup rate at the given conditions would take 1.7 seconds and would also be detectable in the data.

Considering the slow makeup capability of the portable life support system, the slow pressure decay rate of the pressure garment assembly, and the capability to detect, in the data, pressure changes greater than 0.04 psi which last for more than 1 second, there is no evidence that indicates a system malfunction. The crewman had a stuffy head condition during this time period. "Popping" the ears was ruled out, but some other effect internal to the ear may have created the sensation.

This anomaly is closed.

### 14.3.9 Stoppage of the Lunar Surface Camera Counter

The exposure light on the lunar surface close-up camera came on for each exposure, but the mechanical exposure counter did not count every exposed frame. The counter is housed in the handle, which is a matte surface, uncoated aluminum casting. Postflight analysis has indicated that, during extravehicular activity, the camera reached a stabilized handle temperature of approximately 220° F, which is above the mechanical interference point for the counter.

Calculations show that painting the handle white will reduce the stabilization temperature to approximately 110° F, which is a satisfactory operating temperature for the counter. Camera handle castings will be painted white for future missions.

This anomaly is closed.

### 14.3.10 70-mm Lunar-Surface Camera Difficulties

During the second extravehicular period, the Commander's camera did not advance and count every time the trigger was squeezed. Shortly afterwards, when both the camera assemblies were being removed from the remote control units in order to exchange them, both assemblies were loose, although they had been well tightened before egress. In the process of retightening on the lunar surface, the thumbwheel fell off the Lunar Module Pilot's camera assembly, making reassembly impossible (fig. 14-45). The empty camera and faulty assembly were then discarded. The Commander's camera assembly was retightened and performed satisfactorily during the remainder of the extravehicular activity.

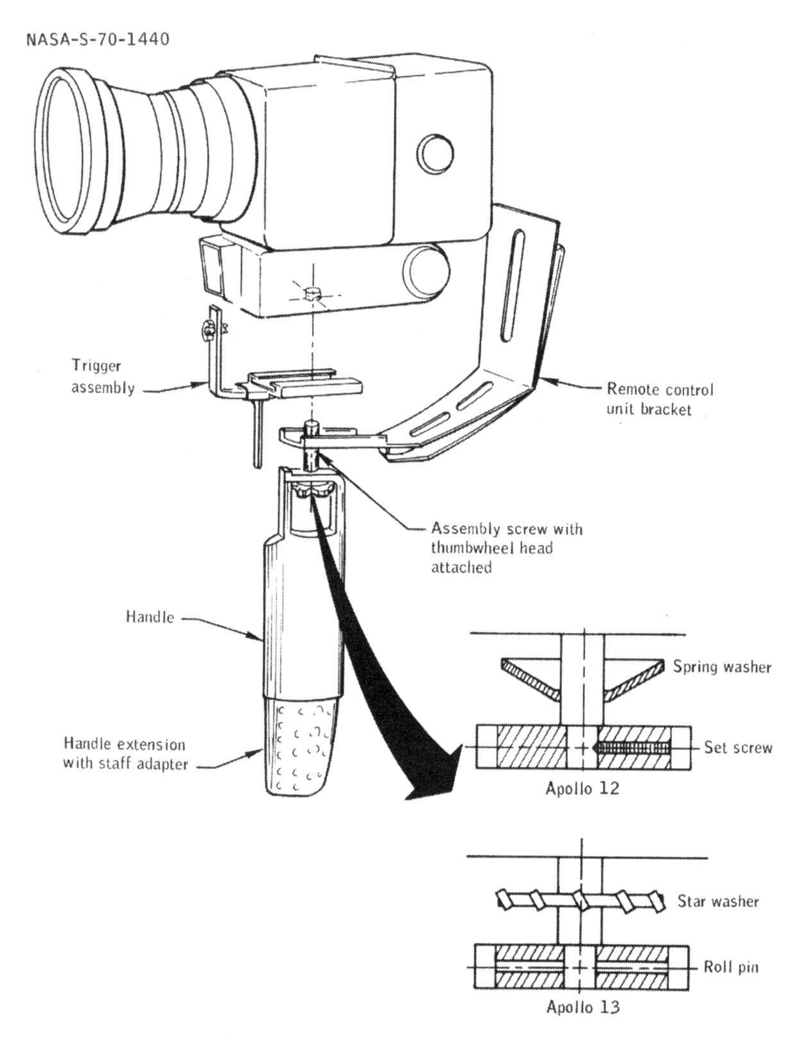


Figure 14-45 70-mm camera handle assembly.

The intermittency experienced by the Commander in the shutter, counter, and film advance actions was the result of excessive trigger play caused by the loose assembly. The loss of the thumbwheel experienced by the Lunar Module Pilot was apparently the result of the improper installation of the thumbwheel setscrew.

For future missions, the cupped spring washer will be replaced by a star washer to resist rotation and loosening of the assembly screw, and the thumbwheel will be secured to the screw with a roll pin, instead of a setscrew.

This anomaly is closed.

### 14.3.11 Tone and Noise During Extravehicular Activity

An undesirable tone, accompanied by a random impulse noise signal, was present intermittently for the first 1-1/2 hours of initial extravehicular activity. The same tone, but without the noise, was present for approximately 12 seconds during the second extravehicular period. This condition did not degrade voice communication but was annoying to the crewmen.

A subsequent analysis of the telemetry data transmitted from the extravehicular mobility unit did not show any degradation of data quality as a result of the noise. Power spectral density plots, however, revealed a fundamental frequency of approximately 1260 hertz and a harmonic frequency of 2520 hertz. Postflight interference tests of an equivalent extravehicular mobility unit revealed the same 1260-hertz tone on the battery bus leads and shield which originated from the fan-motor ripple current. This condition is normal and has been noted during qualification testing of the extravehicular mobility unit. Figure 14-46 illustrates the tone interference generated by the fan motor. However, during these initial tests, the noise interference could not be made to enter the audio system such that the audio tone heard in flight was simulated.

Later laboratory testing of the communications carrier headset demonstrated that lowering a microphone amplifier supply voltage below the regulator threshold of 12.5 volts caused tone interference to enter the audio system. Subsequent analysis showed that a high resistance or the failure of a regulating diode or a transistor in the microphone amplifier regulator could result in a loss of regulator filtering action. The normal operating voltage for the microphone amplifier is from 15.7 to 20.5 volts. When the microphone amplifier supply voltage is above the regulator threshold of 12.5 volts, the tone interference does not enter the audio system.

Postflight tests of the flight communications carriers revealed that the Commander's left microphone was intermittent. Although this failure could not be correlated to the tone phenomena, the random impulse noise heard inflight could be related to the intermittent microphone because a failure analysis has revealed an intermittent open-circuit condition in the primary winding of the amplifier transformer. Additional tests showed no further malfunctions in the communications carriers or harnesses.

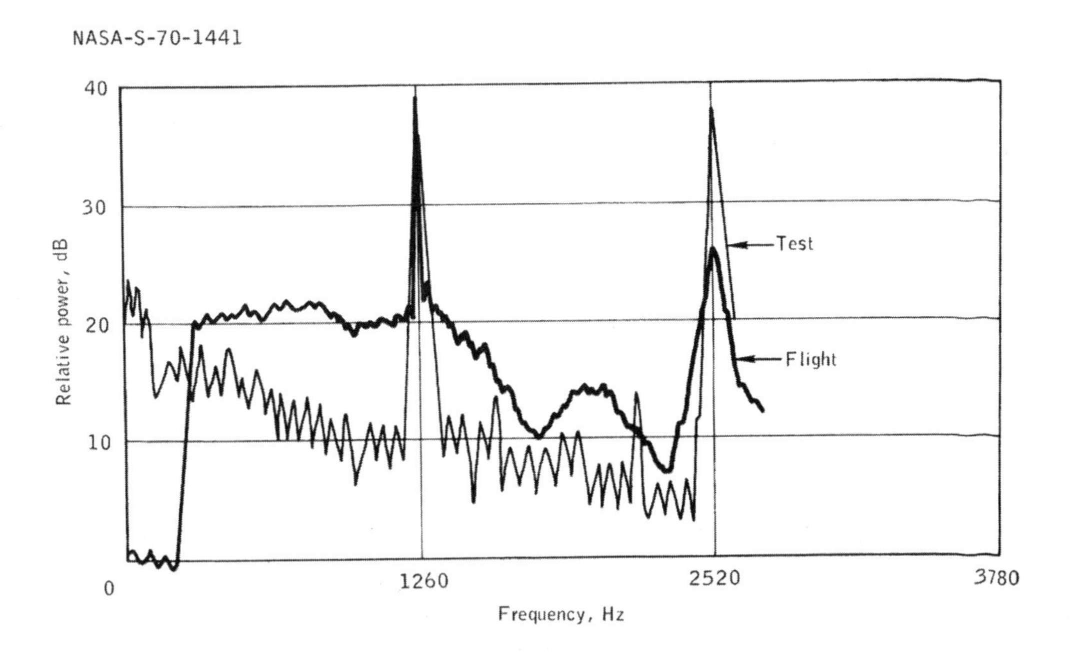


Figure 14-46 Tone power spectral density.

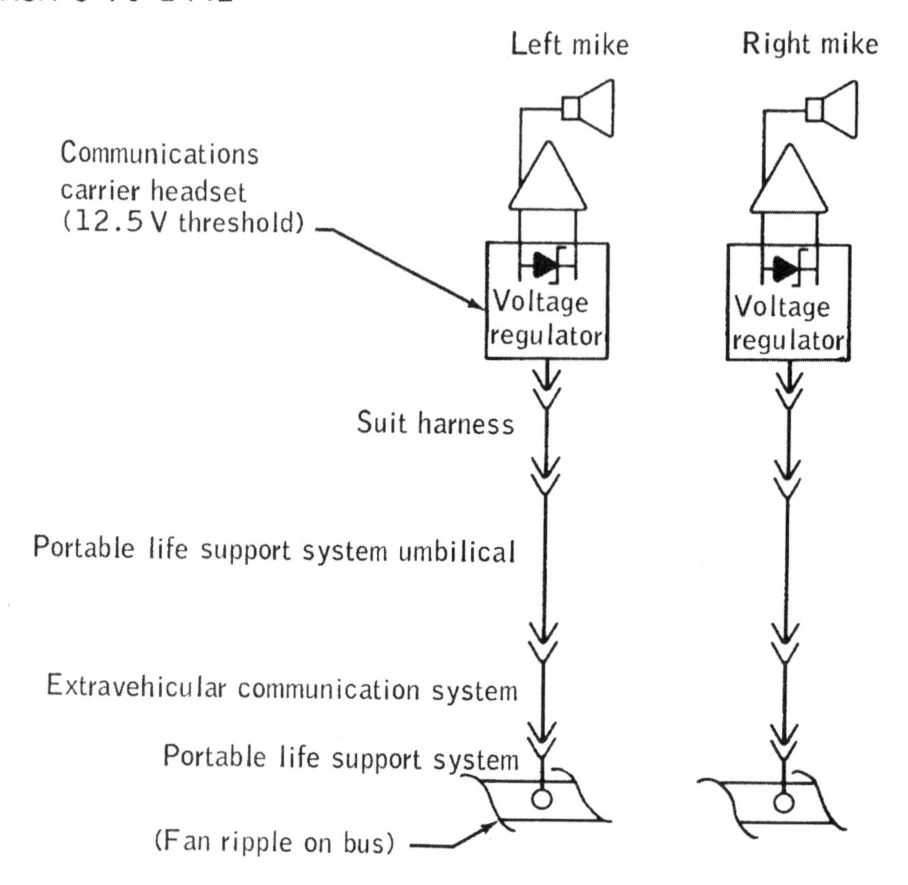


Figure 14-47 Communications carrier headset power path.

#### 14.3.12 Cracked Weigh Bags

The weigh bags were apparently too brittle and therefore cracked and tore when handled on the lunar surface. Those stowed in the sample return container were used to hold the samples of lunar surface material for weighing, and those stowed in the equipment transfer bag were used as collection containers (tote bags) during the geology traverse.

During the traverse, there was a tendency for samples to float out of the bag. Therefore, some means should have been available for opening and closing the bags as required, while maintaining a tight seal when stowed in the spacecraft under zero-g.

The Apollo 12 weigh bags were made from Teflon film. For Apollo 13, the collection containers will be made of a Teflon cloth, which is more flexible and is not as subject to cracks and tears. For Apollo 14 and subsequent missions, both the weigh bags and the collection containers will be constructed from the Teflon cloth. The collection containers will also include a means for repeated opening and closing, as well as providing a tight seal for stowage of return samples in the spacecraft.

This anomaly is closed.

# 15.0 CONCLUSIONS

The Apollo 12 mission demonstrated the capability for performing a precision lunar landing, which is a requirement for the success in future lunar surface explorations. The excellent performance of the spacecraft, the crew, and the supporting ground elements resulted in a wealth of scientific information. The following conclusions are drawn from the information contained in this report.

1. The effectiveness of crew training, flight planning, and realtime navigation from the ground resulted in a precision landing near a previously landed Surveyor spacecraft and well within the desired landing footprint.

2. A hybrid non-free-return translunar profile was flown to demonstrate a capability for additional maneuvering which will be required for future landings to greater latitudes.

3. The timeline activities and metabolic loads associated with the extended lunar surface scientific exploration were within the capability of the crew and the portable life support system.

4. An Apollo lunar surface experiments package was deployed for the first time and, despite some operating anomalies, has returned valuable scientific data in a variety of study areas.

# REFERENCES


1. Journal Geophysics: "The Intrinsic Magnetic Field of the Moon," by C. T. Sonett, D. S. Colburn, and R. G. Currie. Res. 72, 5503. 1967.

2. Jet Propulsion Laboratory: Surveyor III Mission Report Part I Mission Description and Performance. TR32-1177. September 1, 1967.

3. Manned Spacecraft Center: Apollo 9 Mission Report. MSC-PA-R-69-2. May 1969.

4. TRW Systems: Lunar Module Soil Erosion and Visibility Investigations Part I - Summary Report. Report 11176-6060-R0-00. August 13, 1969

5. Marshall Space Flight Center: Saturn V Launch Vehicle Flight Evaluation Report AS-507 Apollo 12 Mission. MPR-SAT-FE-70-1. January 30 , 1970 .

6. NASA Apollo Navigation Working Group: Document No. AN-1.3, Chapter 4. Revision 1 Change 3. November 14, 1969.

7. Manned Spacecraft Center: Apollo 8 Mission Report. MSC-PA-R-69-1. February 1969.

8. Manned Spacecraft Center: Apollo 10 Mission Report. MSC-00126. August 1969.

9. Manned Spacecraft Center: Apollo 11 Mission Report. MSC-00171 November 1969.

10. NASA Headquarters: Apollo Flight Mission Assignments. OMSF M-D MA 500-11 (SE 010-000-1). October 1969.

11. Manned Spacecraft Center: Mission Requirement, H-1 Type Mission (Lunar Landing). SPD9-R-051. July 18, 1969.

12. Marshall Space Flight Center, Kennedy Space Center, Manned Spacecraft Center: Analysis of Apollo 12 Lightning Incident. MSC-01540. February 1970.

# APPENDIX A - VEHICLE DESCRIPTIONS

Very few changes were made to the Apollo 12 space vehicle from the Apollo 11 configuration. The spacecraft/launch vehicle adapter was identical to that for Apollo 11, and the only change to the launch escape system was the incorporation of a more reliable motor igniter. There were no significant changes to the Saturn V launch vehicle. The few changes to the command and service modules and to the lunar module were minor and are discussed in the following paragraphs. A description of lunar surface experiment equipment and a listing of spacecraft mass properties are also presented.

## A.1 COMMAND AND SERVICE MODULES

In the sequential system, wiring was rerouted to preclude a single point failure in the abort system logic. In the service propulsion system, filters were added to prevent contamination of the valve actuation system. Four temperature measurements were added in the instrumentation system to assist in determining spacecraft-to-sun orientation when the guidance system was inoperative. In the water management system, a hydrogen separator was added in the line between the fuel cells and water valve panel. An improved gas separator cartridge was substituted for the unit used in Apollo 11. In the displays and controls system, the service propulsion flange high-temperature caution and warning circuitry, which was no longer required, was removed. The scroll assembly in the entry monitor system was modified to incorporate a more reliable scribe emulsion. In the structural and mechanical systems, the canister for the sea dye marker was mechanically pinned in place to preclude inadvertent actuation, and a single nylon loop was added to replace the command module recovery cable and auxiliary nylon loop.

## A.2 LUNAR MODULE

In the thermal control system, a layer each of Inconel foil and of nickel foil and mesh were added to the landing gear secondary struts to provide additional protection against exhaust plume impingement from the reaction control system; also, a portion of the plume shield was no longer required and was removed from the landing gear deployment truss. The structure was modified in accordance with an organized weight reduction program to decrease weight by reducing the thicknesses of the descent shear webs, ascent stage docking structure, base heat shield, propellant tanks, and oxidizer line. Also, to support higher loads, the ascent propellant tank torus clamp was redesigned and was changed from aluminum to steel.

In the reaction control system, the regulated pressure upper warning level was raised from 205 to 218 psia. In the environmental control system, the accumulator quantity indicator in the suit cooling assembly was modified to improve readability. In the water management section, a redesigned spool was incorporated in the water tank select valve to reduce leakage. Also, a backup measurement was added for descent water pressure.

The following changes were incorporated in the crew provisions as a result of the Apollo 11 experience. Two hammocks were added for increased crew comfort during the lunar-surface stay. The valve, hoses, and large urine bags of the waste management system were replaced with a lighter, less complex system of small urine bags. A condensate collection assembly, having a flow indicator, was added to permit recharging of the water in the portable life support system. The lunar equipment conveyor was redesigned to a single strap arrangement to preclude any possible binding caused by lunar dust. A color television camera was substituted for the slow-scan black-and-white lunar surface camera.

## A.3 EXPERIMENT EQUIPMENT

The Apollo 12 experiment equipment included an Apollo lunar surface experiments package instead of the early Apollo scientific experiments package carried on Apollo 11. The seismic experiments in the two packages were similar in purpose but of different configurations; the other experiments for the Apollo 12 package were new. The solar wind composition experiment and the lunar field geology tools were essentially the same as the Apollo 11 equipment.

The Apollo lunar surface experiments package consists of two subpackages (figs. A-1 and A-2), which were stowed in the lunar module scientific equipment bay for transportation to the moon. In addition the fuel cask containing the radioisotope capsule assembly (part of the electrical power system) was mounted on the external structure of the lunar module. The experiment package includes a central station, an electrical power system, and four experiments: passive seismic, solar wind spectrometer, magnetometer, and suprathermal ion detector. A cold cathode gage is associated with the suprathermal ion detector experiment. The two subpackages could be carried by one man (bar bell arrangement) using the antenna mast as the handle. After the experiments were removed, the subpackage I structure and thermal assembly containing the data subsystem was used as the central station on the lunar surface. The subpackage 2 structure and thermal assembly was used for mounting the electrical power source.

NASA-S-70-1443

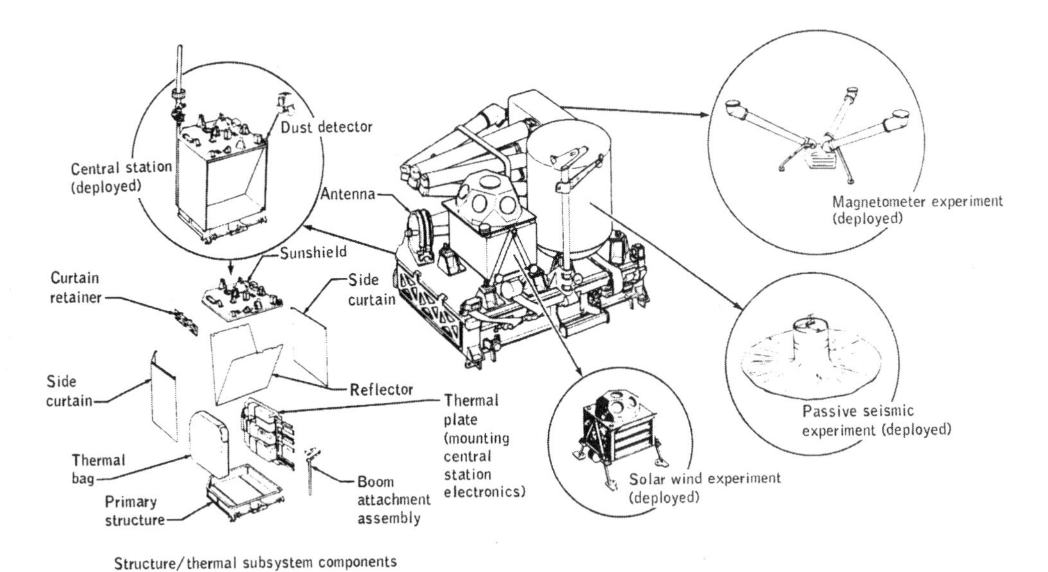


Figure A-1 Experiment subpackage no. 1.

NASA-S-70-1444

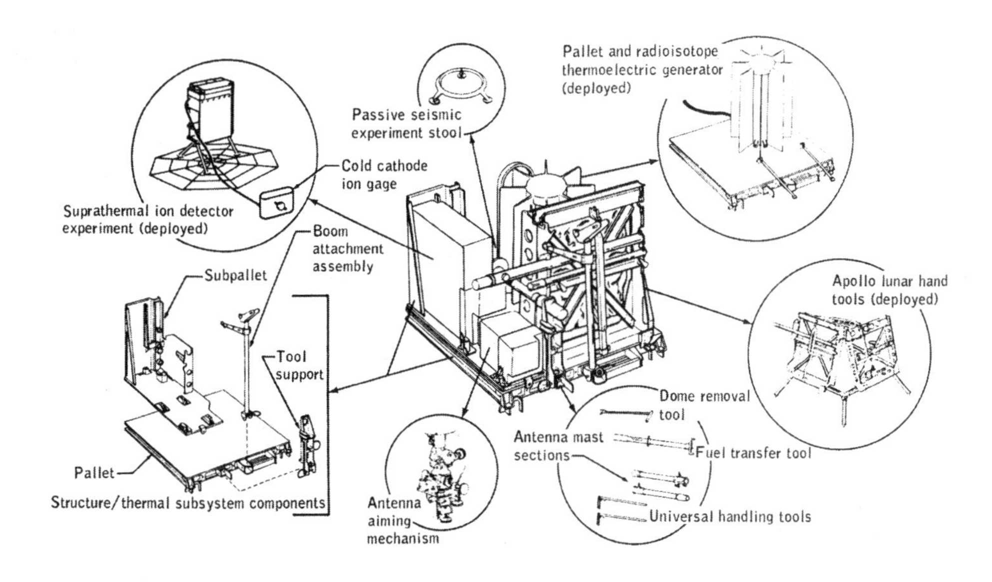


Figure A-2 Experiment subpackage no. 2.

### A.3.1 Central Station

The central station (fig. A-1) is the focal point for control of the experiments and for the collection, processing, and transmission of scientific and engineering data to the Manned Space Flight Network.

The central station includes a data system consisting of an antenna, a diplexer, transmitter, command receiver and decoder, timer, data processor, and power distribution unit.

The antenna, consisting of a copper conductor bonded to a fiberglass epoxy tube for mechanical support, is a modified axial helix capable of receiving and transmitting a right-hand circularly polarized S-band signal. A two-gimbal aiming mechanism permits the position of the antenna to be adjusted in azimuth and elevation. The diplexer consists of a filter that provides the attenuation required at the operating frequencies and a circulator switch that couples the selected transmitter (A or B) to the antenna. Two mutually redundant transmitters generate an S-band carrier frequency between 2275 and 2280 megahertz. The carrier is phase modulated by the bit stream from the data processor. The command receiver receives the uplink commands transmitted from the earth stations. The command decoder provides the digital timing and command data and applies the commands required to control the operation of the experiments. The timer provides predetermined switch closures to initiate specific functions within the experiments and data system when the uplink commands are not available. The timer consists of a clock and a long life mercury cell battery. The data processor includes two mutually redundant data processing channels, each of which generates experiment timing and control signals, collects and formats experiment data, and provides data for phase modulation of the transmitted carrier. The power distribution unit contains the circuitry for the power-off sequencer, monitors temperature and voltage, and controls power for experiments and central station.

A dust detector mounted on the central station measures the dust accumulation. The detector consists of a sensor, which has three photo cells, and associated circuitry.

### A.3.2 Electrical Power System

The electrical power system (fig. A-2) provides the power for operation of the experiment packages. The primary electrical energy is developed by thermoelectric action with thermal energy supplied by a radioisotope source. The expected output is a constant 16 volts.

The electrical power system consists of a radioisotope thermoelectric generator, fuel capsule assembly, power conditioning unit, and fuel cask. The radioisotope thermoelectric generator is a cylindrical case with eight heat rejection fins on the exterior and an interior thermo pile to receive the fuel capsule. The fuel capsule is a thin-walled cylindrical structure containing the radioisotope fuel, plutonium 238. The power conditioning unit contains the dc voltage converters, shunt regulators, filters, and amplifiers required to convert and regulate the power. The graphite fuel cask, a cylindrical structure with a threaded cover, was used to transport the fuel capsule from the earth to the moon.

### A.3.3 Passive Seismic Experiment

The passive seismic experiment (fig. A-1) monitors seismic activity and detects meteoroid impacts and free oscillations. It also detects surface tilt produced by tidal deformations resulting, in part, from periodic variations in the strength and direction of external gravitational fields acting on the moon and from changes in the vertical component of gravitational acceleration.

The experiment consists of a sensor assembly, leveling stool, thermal shroud, and an electronics assembly. The sensor assembly contains one vertical short period seismometer and three orthogonally aligned long period seismometers. The leveling stool is a short tripod that holds the sensor and permitted the crewman to level the sensor to within 5 degrees of vertical. The stool also provides thermal and electrical insulation of the sensor from the lunar surface but at the same time can transmit surface motion having frequencies of up to 26.5 hertz, with negligible attenuation. The thermal shroud consists of 10 layers of aluminized Mylar separated by alternate layers of silk cord wound on a perforated aluminum support. The shroud aids in stabilizing the temperature of the sensor assembly.

The electronics assembly is functionally a part of the passive seismic experiment but is physically a part of the central station. The electronics assembly contains circuitry associated with the attenuating, amplifying, and filtering of the seismic signals, processing of the applicable data, and the internal power supplies.

### A.3.4 Solar Wind Spectrometer

The solar wind spectrometer (fig. A-1) measures energies, densities, incidence angles, and temporal variations of the electron and proton components of the solar wind plasma that strikes the lunar surface.

The experiment consists of a sensor assembly, electronic assembly, thermal control assembly, and leg assembly. The sensor assembly contains seven Faraday cups, which measure the current produced by the charged particle flux that enters. The electronic assembly contains the circuitry for modulating the plasma flux entering the Faraday cups and for converting the data into a digital format appropriate for the central station. The thermal control assembly includes three radiators on one vertical face and insulation on the outer faces of the electronic assembly. The leg assembly consists of two tubular A-frames containing telescoping legs.

### A.3.5 Magnetometer

The magnetometer (fig. A-1) measures the magnetic fields resulting from internal and external lunar forces to provide some indication of the composition of the lunar interior.

The experiment consists of three magnetic (flux-gate) sensors mounted on the ends of orthogonal 3-foot support arms. The support arms extend from an electronics and gimbal-flip unit, which is enclosed by a fiberglass protective cover underneath a thermal blanket. The sensors are wrapped with insulation, except for their upper flat surfaces, which serve as heat radiators. Leveling legs are attached to the base of each support arm.

### A.3.6 Suprathermal Ion Detector

The suprathermal ion detector experiment (fig. A-2) measures the ions streaming from the ultraviolet ionization of the lunar atmosphere and from the solar wind. The cold cathode gage measures the density of the lunar atmosphere.

The suprathermal ion detector consists of two curved plate analyzers and a ground plane. One analyzer counts the low energy ions (velocity range of 40 000 to 9 350 000 cm/sec and energy range of 0.2 to 48.6 electron volts). The other analyzer counts the high energy ions at selected energy intervals between 10 and 3500 electron volts. The electrical potential between the analyzers and the lunar surface is controlled by applying a known voltage between the analyzers and the ground plane. The cold cathode gage determines the pressure of the ambient lunar atmosphere over the range of $10^{-6}$ to $10^{-12}$ torr.

### A.4 MASS PROPERTIES

Spacecraft mass properties for the Apollo 12 mission are summarized in table A-I. These data represent the conditions as determined from postflight analyses of expendable loadings and usage during the flight. Variations in spacecraft mass properties are determined for each significant mission phase from lift-off through landing. Expendables usage is based on reported real-time and postflight data as presented in other sections of this report. The weights and centers of gravity of the individual command and service modules and of the lunar module ascent and descent stages were measured prior to flight, and the inertia values were calculated. All changes incorporated after the actual weighing were monitored, and the spacecraft mass properties were updated.

TABLE A-I.- MASS PROPERTIES

| Event | Weight, lb | Center of gravity, in. | | | Moment of inertia, slug-ft$^2$ | | | Product of inertia, slug-ft$^2$ | | |
|---|---|---|---|---|---|---|---|---|---|---|
| | | $X_A$ | $Y_A$ | $Z_A$ | $I_{XX}$ | $I_{YY}$ | $I_{ZZ}$ | $I_{XY}$ | $I_{XZ}$ | $I_{YZ}$ |
| Lift-off | 110 090.3 | 846.6 | 2.4 | 3.8 | 67 785 | 1 173 398 | 1 175 941 | 3055 | 9 618 | 3672 |
| Earth orbit insertion | 101 126.9 | 806.6 | 2.5 | 4.1 | 66 935 | 717 363 | 719 955 | 4955 | 12 028 | 3357 |
| Transposition and docking | | | | | | | | | | |
| Command & service modules | 63 535.6 | 934.1 | 4.0 | 6.5 | 33 931 | 75 941 | 78 546 | -1837 | -66 | 3179 |
| Lunar module | 33 584.2 | 1237.0 | -.2 | .0 | 22 540 | 24 713 | 25 252 | -455 | 94 | 274 |
| Total docked | 97 119.8 | 1038.9 | 2.5 | 4.3 | 56 753 | 535 814 | 538 840 | -8258 | -9285 | 3581 |
| First midcourse correction | | | | | | | | | | |
| Ignition | 96 870.6 | 1039.1 | 2.6 | 4.3 | 56 534 | 534 840 | 537 907 | -8313 | -9232 | 3643 |
| Cutoff | 96 401.2 | 1039.5 | 2.5 | 4.2 | 56 289 | 534 105 | 537 375 | -8307 | -9181 | 3575 |
| Lunar orbit insertion | | | | | | | | | | |
| Ignition | 96 261.1 | 1039.6 | 2.6 | 4.2 | 56 201 | 533 591 | 536 872 | -8393 | -9079 | 3609 |
| Cutoff | 72 335.6 | 1080.2 | 1.5 | 2.9 | 43 798 | 414 533 | 421 908 | -6191 | -5179 | 686 |
| Circularization | | | | | | | | | | |
| Ignition | 72 243.7 | 1080.4 | 1.6 | 2.9 | 43 711 | 414 139 | 421 538 | -6209 | -5154 | 708 |
| Cutoff | 71 028.4 | 1082.9 | 1.4 | 2.9 | 43 096 | 408 156 | 414 962 | -5823 | -5207 | 633 |
| Separation | 70 897.3 | 1083.9 | 1.6 | 2.8 | 44 317 | 408 272 | 415 121 | -5487 | -5416 | 611 |
| Docking | | | | | | | | | | |
| Command & service modules | 35 306.2 | 944.7 | 2.5 | 5.7 | 19 345 | 55 835 | 61 584 | -2083 | 829 | 326 |
| Ascent stage | 5 765.6 | 1168.7 | 4.3 | -2.0 | 3 341 | 2 361 | 2 680 | -146 | 17 | -285 |
| Total after docking | | | | | | | | | | |
| Ascent stage manned | 41 071.8 | 976.1 | 2.7 | 4.6 | 22 752 | 111 934 | 117 943 | -1794 | -989 | 25 |
| Ascent stage unmanned | 41 059.4 | 974.6 | 2.5 | 4.6 | 22 652 | 108 717 | 114 655 | -2278 | -756 | 60 |
| After ascent stage jettison | 35 622.9 | 945.0 | 2.6 | 5.5 | 19 432 | 55 624 | 61 357 | -2012 | 700 | 322 |
| Transearth injection | | | | | | | | | | |
| Ignition | 34 130.6 | 946.2 | 2.4 | 5.6 | 18 576 | 55 260 | 60 417 | -1916 | 691 | 300 |
| Cutoff | 25 724.5 | 965.5 | -.5 | 6.9 | 14 268 | 46 636 | 47 715 | -646 | 115 | -160 |
| Command and service module separation | | | | | | | | | | |
| Before | 25 444.2 | 966.0 | -.4 | 6.8 | 14 057 | 46 417 | 47 515 | -685 | 177 | -100 |
| After | | | | | | | | | | |
| Service module | 13 160.7 | 897.0 | -.4 | 7.5 | 8 250 | 13 492 | 15 134 | -773 | 870 | -101 |
| Command module | 12 283.5 | 1039.9 | -.3 | 6.1 | 5 803 | 4 934 | 4 393 | 66 | -401 | 0 |
| Entry | 12 275.5 | 1039.9 | -.3 | 6.0 | 5 799 | 4 930 | 4 392 | 66 | -400 | 0 |
| Drogue deployment | 11 785.7 | 1038.6 | -.3 | 6.0 | 5 612 | 4 596 | 4 084 | 67 | -375 | 0 |
| Main parachute deployment | 11 496.1 | 1038.5 | -.2 | 5.3 | 5 471 | 4 406 | 4 021 | 61 | -312 | 18 |
| Landing | 11 050.2 | 1036.5 | -.2 | 5.2 | 5 405 | 4 123 | 3 720 | 55 | -321 | 18 |
| Lunar Module | | | | | | | | | | |
| Lunar module at launch | 33 586.9 | 185.3 | -.0 | -.2 | 22 545 | 24 837 | 25 047 | 150 | 440 | 368 |
| Separation | 33 985.5 | 186.2 | -.0 | .4 | 23 908 | 25 928 | 26 009 | 149 | 685 | 366 |
| Descent orbit insertion | | | | | | | | | | |
| Ignition | 33 971.8 | 186.2 | -.0 | .4 | 23 899 | 25 911 | 25 989 | 148 | 684 | 363 |
| Cutoff | 33 719.3 | 186.2 | -.0 | .4 | 23 740 | 25 849 | 25 964 | 148 | 684 | 363 |
| Lunar landing | 16 564.2 | 211.0 | -.0 | .8 | 12 921 | 14 441 | 16 981 | 147 | 612 | 366 |
| Lunar lift-off | 10 749.6 | 243.6 | .2 | 2.5 | 6 727 | 3 263 | 5 936 | 67 | 196 | -10 |
| Orbit insertion | 5 965.6 | 255.0 | .4 | 4.6 | 3 430 | 2 893 | 2 307 | 57 | 129 | -11 |
| Coelliptic sequence initiation | 5 885.9 | 254.6 | .4 | 4.6 | 3 394 | 2 874 | 2 260 | 57 | 131 | -8 |
| Docking | 5 765.6 | 254.1 | .4 | 4.7 | 3 341 | 2 848 | 2 193 | 57 | 135 | -4 |
| Jettison | 5 436.5 | 254.7 | .2 | 2.2 | 3 178 | 2 816 | 2 233 | 76 | 120 | -26 |

# APPENDIX B - SPACECRAFT HISTORIES

The history of command and service module (CSM 108) operations at the manufacturer's facility, Downey, California, is shown in figure B-1, and the operations at Kennedy Space Center, Florida, in figure B-2.

The history of the lunar module (LM-6) at the manufacturer's facility, Bethpage, New York, is shown in figure B-3, and the operations at Kennedy Space Center, Florida, in figure B-4.

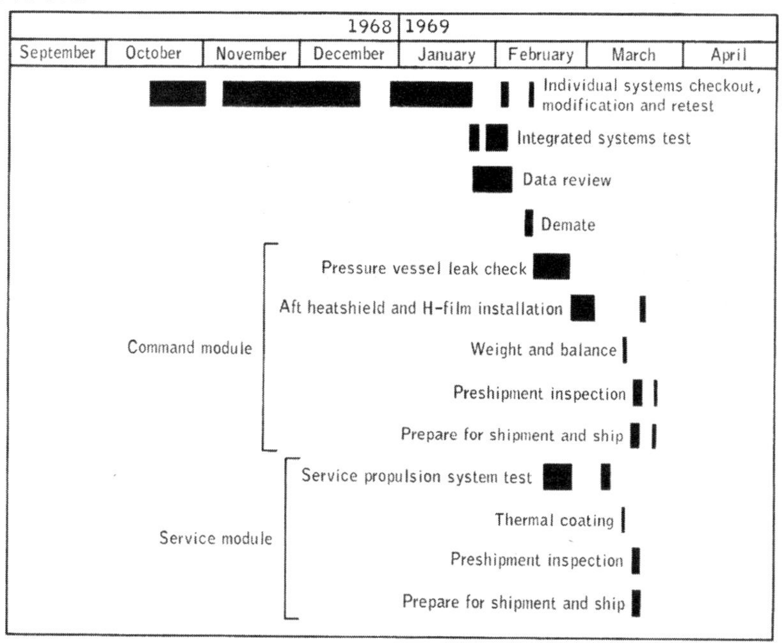


Figure B-1 Factory checkout flow for the command and service modules at Contractor's facility.

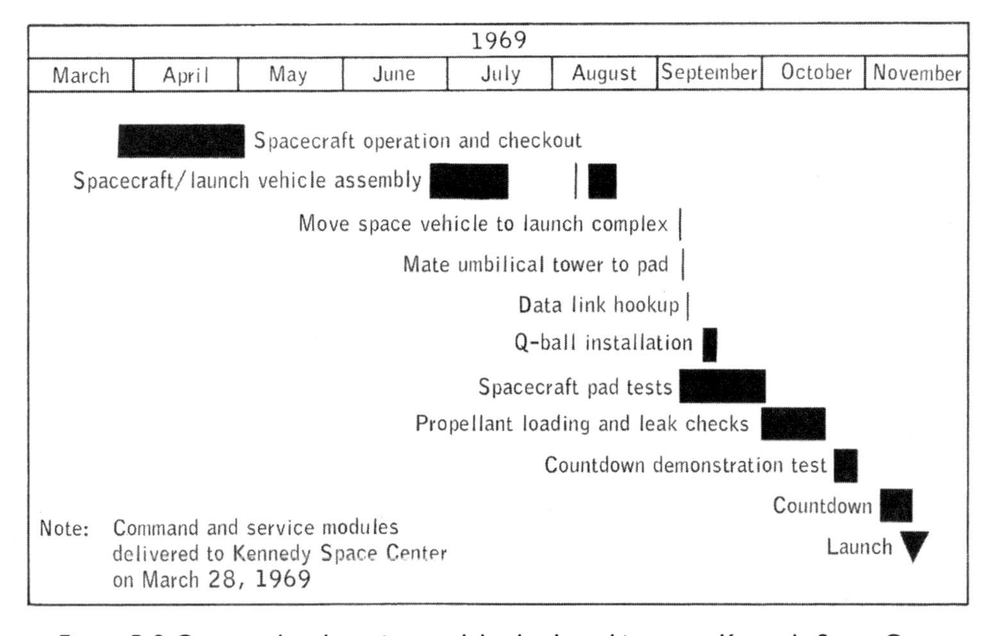


Figure B-2 Command and service module checkout history at Kennedy Space Center.

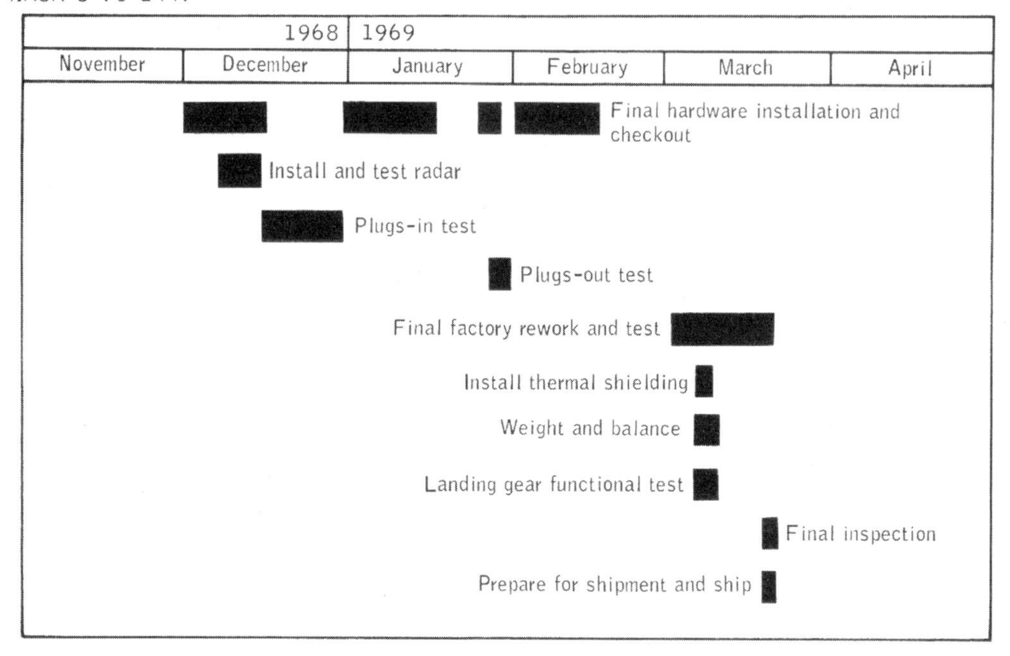


Figure B-3 Factory checkout flow for the lunar module at Contractor's facility.

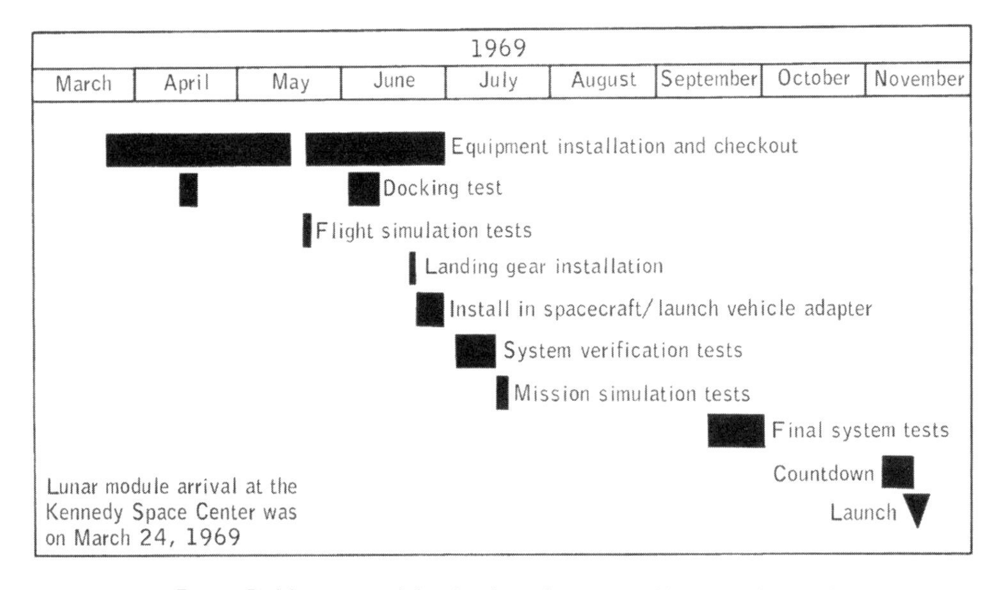


Figure B-4 Lunar module checkout history at Kennedy Space Center.

# APPENDIX C - POSTFLIGHT TESTING

The command module arrived at the Lunar Receiving Laboratory, Houston, Texas, on December 2, 1969, after reaction control system deactivation and pyrotechnic safing in Hawaii. At the end of the quarantine period, the command module was shipped to the contractor's facility in Downey, California, on January 11. Postflight testing and inspection of the command module for evaluation of the inflight performance and investigation of the flight irregularities were conducted at the contractor's and vendor's facilities and at the Manned Spacecraft Center in accordance with approved Apollo Spacecraft Hardware Utilization Requests (ASHUR's). The tests performed as a result of inflight problems are described in table C-I and discussed in the appropriate systems performance sections of this report. Tests being conducted for other purposes in accordance with other ASHUR's and the basic contract are not included.

TABLE C-I.- POSTFLIGHT TESTING SUMMARY

| ASHUR no. | Purpose | Tests performed | Results |
|---|---|---|---|
| | | Displays and Controls | |
| 108021 | To determine the cause of the intermittent tuning fork display indication on the panel 2 mission clock. | Determine solder joint integrity and wiring continuity. Perform failure analysis. | Continuity check satisfactory. Unable to duplicate failure. |
| | | Guidance and Navigation | |
| 108008 | To investigate the cause for optics coupling display unit indication of optics movement during the zero optics mode. | Perform operational test. | Not complete. |
| | | Electrical Power | |
| 108023 | To determine why circuit breaker (CB23) was open during earth orbit checks. | Perform pull test, mounting torque test, and calibration check. | CB23 normal mechanically and electrically. |
| | | Communications | |
| 108002 | To determine the cause for the failure of the color television. | Perform failure analysis. | Potassium chloride burned off the target. |
| 108019<br>108020 | To investigate the extravehicular activity tone problem. | Perform functional test of communication carriers and bioinstrumentation. | Tone was duplicated by lowering the voltage at the microphone. |
| 108022<br>108035 | To determine the cause of the VHF garbled voice. | Perform functional and systems tests of the VHF/AM transceiver, audio center, digital ranging generator, and lightweight headset. | VHF intelligibility dependent on range and squelch setting. Also dependent on lightweight headset microphone placement. |
| 108054 | To investigate the failure of two VHF ground plane radials. | Inspect and actuate the VHF ground plane radials. | Ground plane radials deployment fouled by canvas flap. |
| | | Environmental Control | |
| 108004 | To investigate the unexplained high oxygen use rate. | Determine the pressure integrity of the oxygen lines and tanks. | No leakage in the command module portion of the system. |
| 108005 | To investigate the plugged urine filters. | Determine water flow rate and pressure drop. Disassemble to determine quantity and source of contaminants. | Plugging caused by urine solids. |
| 108006 | To investigate the shift in the suit pressure transducer. | Calibrate the transducer. Perform failure analysis. | Calibration verified shift. Analysis not complete. |

TABLE C-I.- POSTFLIGHT TESTING SUMMARY - Concluded

| AHSUR no. | Purpose | Tests performed | Results |
|---|---|---|---|
| | | Environmental Control - concluded | |
| 108029 | To investigate the leak at the food preparation water port. | Measure the water leak. Perform teardown and analysis. | Unit did not leak with unheated water. Leak duplicated with heated water. Analysis not complete. |
| 108039<br>108049<br>108050<br>108058 | To investigate the excessive quantity of particulate matter in the command module cabin. | Perform material analysis of particles in the lithium hydroxide cartridge, oxygen umbilicals, environmental control system ducts and filters. | Not complete. |
| 108053 | To determine the cause for erratic operation of the potable water transducer. | Perform failure analysis. | Not complete. |
| | | Crew Equipment | |
| 108026 | To investigate reported fraying of the exercise rope. | Perform visual inspection. | Rope showed normal wear. |
| 108028 | To investigate reported difficulties with the tape cassettes and voice recorder. | Perform operational tests, disassembly, and inspection. | Recorder performance satisfactory. |
| 108034 | To investigate intermittent operation of of 16-mm camera. | Perform inspection and performance checks. | Camera performed satisfactorily. Film perforations did not line up with index line on front of two magazines. |
| 108051 | To investigate possible failure of 70-mm camera. | Perform visual inspection and operational tests. | One magazine performed satisfactorily. One magazine of infrared film showed lines caused by normal heat from film rollers (lines were between frames). |

# APPENDIX D - DATA AVAILABILITY

Tables D-I and D-II are summaries of the data made available for systems performance analyses and anomaly investigations. Table D-I lists the data from the command and service modules, and table D-II, the lunar module. For additional information regarding data availability, the status listing of all mission data in the Central Metric Data File, building 12, MSC, should be consulted.

TABLE D-I.- COMMAND AND SERVICE MODULE DATA AVAILABILITY

| Time, hr:min | | Range station | Bandpass plots or tabs | Bilevels | Computer words | O'graph records | Brush records | Special plots or tabs | Special programs |
|---|---|---|---|---|---|---|---|---|---|
| From | To | | | | | | | | |
| -04:00 | +00:02 | ALDS | X | | | | | | |
| -00:02 | 00:03 | GDS[a] | X | X | X | X | X | X | X |
| -00:01 | 00:12 | MILA | X | X | X | X | X | X | X |
| 00:00 | 03:34 | MSFN | X | | X | | | | |
| +00:02 | 00:14 | BDA | X | | | X | | X | |
| 00:07 | 00:18 | VAN | X | X | | | | | |
| 00:25 | 00:53 | VAN[a] | X | | | | | X | |
| 01:03 | 01:29 | VAN[a] | X | | | | | X | |
| 01:55 | 02:44 | MAD[a] | X | | | | X | | |
| 02:42 | 02:54 | ARIA[a] | X | | X | X | X | X | |
| 02:45 | 03:40 | GDS | | | | | | | X |
| 02:45 | 02:52 | MAD[a] | X | | X | X | X | | |
| 02:48 | 03:05 | HAW | X | | X | X | X | X | |
| 02:54 | 83:11 | MSFN | | | | | | | X |
| 03:13 | 03:31 | GDS | X | X | X | X | | X | X |
| 03:34 | 08:37 | MSFN | X | | X | | | | |
| 03:54 | 04:07 | GDS | | | | | | | X |
| 04:08 | 04:24 | GDS | X | X | | X | | X | X |
| 04:43 | 05:12 | GDS | | | | | | | X |
| 08:37 | 11:29 | MSFN | X | | X | | | | |
| 10:49 | 10:52 | GDS | | | | | X | | |
| 11:29 | 15:25 | MSFN | X | | X | | | | |
| 16:22 | 31:39 | MSFN | X | | X | | | | |
| 29:42 | 30:41 | GDS | | | | | | | X |
| 30:35 | 31:05 | MAD | | | | | | | X |
| 30:40 | 31:09 | GDS | X | X | | X | X | X | X |
| 30:50 | 31:00 | GDS | | | X | X | | | |
| 31:00 | 32:03 | GDS | | | | | X | | |
| 31:27 | 31:45 | GDS | X | X | | | X | | |
| 31:39 | 31:44 | MSFN | | X | | | | | |
| 31:39 | 39:40 | MSFN | X | | X | | | | |
| 35:39 | 35:46 | GDS | | | | | X | | |
| 38:01 | 43:31 | MSFN | X | | X | | | | |
| 39:25 | 39:36 | GDS | | | | | | | X |
| 41:19 | 41:21 | HSK | | | | | X | | |
| 43:38 | 59:30 | MSFN | X | | X | | | | |
| 54:12 | 54:20 | GDS | | | | | | | X |
| 57:39 | 57:41 | GDS | | | | | X | | |
| 59:30 | 67:21 | MSFN | X | | X | | | | |
| 62:54 | 63:15 | GDS | | | | | X | | X |
| 64:04 | 64:12 | GDS | | | | | | | X |
| 67:21 | 83:11 | MSFN | X | | X | | | | |
| 83:11 | 83:23 | GDS[a] | X | | X | | X | X | |
| 83:11 | 87:12 | MSFN | X | | X | | | | |
| 83:23 | 83:33 | GDS[a] | X | X | X | X | X | X | |
| 83:33 | 83:44 | GDS[a] | X | | X | | X | | |
| 84:10 | 84:45 | MSFN | | | | | | X | |
| 84:15 | 85:10 | GDS | | | | | | | X |
| 85:11 | 85:52 | GDS[a] | X | | | | | X | |
| 86:50 | 87:00 | GDS | | | | | | | X |
| 87:12 | 91:11 | MSFN | X | | X | | | | |
| 87:17 | 88:01 | HSK[a] | X | | | | | X | |
| 87:46 | 87:51 | HSK[a] | X | X | X | X | | X | |
| 89:13 | 90:47 | HSK[a] | X | | | | | X | |
| 90:40 | 91:11 | HSK | | | | | | | X |
| 91:07 | 95:07 | MSFN | X | | X | | | | |

[a]Data dump

TABLE D-I.- COMMAND AND SERVICE MODULE DATA AVAILABILITY - Continued

| Time, hr:min | | Range station | Bandpass plots or tabs | Bilevels | Computer words | O'graph records | Brush records | Special plots or tabs | Special programs |
|---|---|---|---|---|---|---|---|---|---|
| From | To | | | | | | | | |
| 91:11 | 91:58 | HSK[a] | X | | | | | X | |
| 93:09 | 93:56 | HSK[a] | X | | | | | X | |
| 95:07 | 95:54 | MAD[a] | X | | | | | X | |
| 95:07 | 98:35 | MSFN | X | | X | | | | |
| 97:05 | 97:53 | MAD[a] | X | | | | | X | |
| 97:50 | 98:40 | MAD | | | | | | | X |
| 98:35 | 102:53 | MSFN | X | | X | | | | |
| 99:04 | 99:52 | MAD[a] | X | | | | | X | |
| 99:57 | 100:57 | MAD | | | | X | | | X |
| 100:40 | 101:10 | MAD | | | | | | | X |
| 100:58 | 101:50 | MAD[a] | X | | | | | X | |
| 102:53 | 106:40 | MSFN | X | | X | | | | |
| 103:00 | 103:48 | GDS[a] | X | | | | | X | |
| 103:51 | 104:01 | GDS | | | | | | | X |
| 104:59 | 105:48 | GDS[a] | X | | | | | X | |
| 106:12 | 106:48 | GDS | | | X | | | | |
| 106:40 | 111:20 | MSFN | X | | X | | | | |
| 107:46 | 108:57 | GDS | | | | | | | X |
| 107:50 | 108:00 | GDS | | X | X | X | | | |
| 108:20 | 108:30 | GDS | | X | X | | | | |
| 108:23 | 108:26 | GDS | | | | X | | | |
| 108:55 | 109:44 | GDS[a] | X | | X | | | X | |
| 109:41 | 110:20 | GDS | | | X | | | | |
| 110:40 | 110:55 | HSK | | | | | | | X |
| 110:54 | 111:54 | GDS[a] | X | | | | | X | |
| 111:20 | 115:39 | MSFN | X | | X | | | | |
| 111:50 | 112:00 | HSK | | | | | | | X |
| 112:03 | 112:30 | GDS | | | X | | | | |
| 114:10 | 114:30 | HSK | | | X | | | | |
| 114:50 | 115:38 | HSK | X | | | | | X | |
| 115:41 | 118:57 | MSFN | X | | X | | | | |
| 115:45 | 116:05 | HSK | | | | | | | X |
| 116:00 | 116:36 | HSK | | | X | | | | |
| 116:49 | 117:30 | HSK[a] | X | | | | | X | |
| 118:46 | 119:35 | MAD[a] | X | | | | | X | |
| 119:17 | 123:06 | MSFN | X | | X | | | | |
| 119:39 | 119:56 | MAD | | | X | | | | |
| 119:43 | 119:58 | MAD | | X | | X | | X | |
| 120:00 | 120:30 | MAD | | | | | | | X |
| 120:30 | 120:36 | MAD | | | | | X | | |
| 120:53 | 121:33 | GDS[a]X[b] | X | | | | | X | |
| 123:06 | 127:40 | MSFN | X | | X | | | | |
| 125:03 | 125:31 | GDS[a] | X | | | | | X | |
| 126:43 | 127:29 | GDS[a]X[b] | X | | | | | X | |
| 127:41 | 131:44 | MSFN | X | | X | | | | |
| 128:39 | 129:29 | GDS[a]X[b] | X | | | | | X | |
| 130:35 | 131:26 | GDS[a] | X | | | | | X | |
| 131:44 | 135:39 | MSFN | X | | X | | | | |
| 132:37 | 133:26 | GDS[a] | X | | | | | X | |
| 133:24 | 134:26 | GDS | | | | | | | X |
| 134:00 | 134:35 | GDS | | | X | | | | |
| 134:35 | 135:22 | HSK[a] | X | | | | | X | |
| 135:39 | 139:20 | MSFN | X | | X | | | | |
| 135:50 | 136:10 | GDS | | | | | | | X |
| 136:33 | 137:21 | HSK[a] | X | | | | | X | |

[a]Data dump

[b]Indicates wing site.

TABLE D-I.- COMMAND AND SERVICE MODULE DATA AVAILABILITY - Continued

| Time, hr:min | | Range station | Bandpass plots or tabs | Bilevels | Computer words | O'graph records | Brush records | Special plots or tabs | Special programs |
|---|---|---|---|---|---|---|---|---|---|
| From | To | | | | | | | | |
| 138:31 | 139:19 | HSK[a] | X | | | | | X | |
| 139:31 | 143:30 | MSFN | X | | X | | | | |
| 140:33 | 141:18 | HSK[a] | X | | | | | X | |
| 142:03 | 142:28 | HSK | | | X | | | | |
| 142:28 | 143:17 | HSK[a] | X | | X | | | X | |
| 143:40 | 147:28 | MSFN | X | | X | | | | |
| 144:26 | 145:09 | MAD[a] | X | | | | | X | |
| 145:35 | 145:38 | MAD | | | | X | | | |
| 146:20 | 146:30 | MAD | | | | | | | X |
| 146:25 | 147:15 | MAD[a] | X | | | | | X | |
| 147:10 | 148:20 | MAD | | | | | | | X |
| 147:28 | 150:06 | MSFN | X | | X | | | | |
| 147:58 | 148:06 | MAD | | X | | X | | | |
| 148:23 | 149:09 | MAD | X | | | | | X | |
| 150:06 | 159:56 | MSFN | X | | X | | | | |
| 156:17 | 157:05 | GDS[a]X[b] | X | | | | | X | |
| 157:20 | 158:20 | GDS | | | | | | | X |
| 158:09 | 158:20 | HSK[a] | X | | X | | | X | |
| 159:01 | 159:10 | HSK | | X | X | X | X | X | |
| 159:04 | 159:20 | GDS | | | X | | | | |
| 159:56 | 163:44 | MSFN | X | | X | | | | |
| 160:02 | 160:11 | GDS | | | | | X | | |
| 162:14 | 162:57 | HSK[a] | X | | | | | X | |
| 163:30 | 163:45 | HSK | | | | | | | X |
| 163:44 | 167:24 | MSFN | X | | X | | | | |
| 164:11 | 165:07 | HSK[a] | X | | | | | X | |
| 165:00 | 165:35 | HSK | | | X | | | | |
| 166:10 | 167:17 | HSK[a] | X | | | | | X | |
| 167:24 | 170:05 | MSFN | X | | X | | | | |
| 168:08 | 168:56 | MAD[a] | X | | X | | | X | |
| 169:20 | 169:30 | MAD | | | X | | | | |
| 170:05 | 175:37 | MSFN | X | | | | | X | |
| 170:06 | 170:58 | MAD[a] | X | | | | | X | |
| 172:25 | 172:32 | MAD[a] | X | X | X | X | | | |
| 172:32 | 172:41 | MAD[a] | X | | | | | X | |
| 172:40 | 244:21 | MSFN | | | | | | | X |
| 173:10 | 173:50 | MAD | | | | | | | X |
| 175:37 | 191:36 | MSFN | X | | X | | | | |
| 188:20 | 188:33 | HSK | X | X | X | X | | | X |
| 189:10 | 189:32 | HSK | | | X | | | | |
| 191:36 | 195:32 | MSFN | X | | X | | | | |
| 192:30 | 194:30 | MAD | | | | | | | X |
| 195:32 | 203:39 | MSFN | X | | X | | | | |
| 200:02 | 200:07 | GDS | | X | | | | | |
| 203:39 | 207:39 | MSFN | X | | X | | | | |
| 205:57 | 206:04 | GDS | | X | | | | | |
| 207:39 | 215:21 | MSFN | X | | X | | | | |
| 212:02 | 212:07 | GDS | | X | | | | | |
| 215:06 | 215:22 | HSK | | | | | X | | |
| 215:21 | 219:35 | MSFN | X | | X | | | | |
| 215:40 | 216:50 | HSK | | | | | | | X |
| 216:00 | 216:27 | HSK | | | | | X | | |
| 218:10 | 219:50 | MAD | | | | | | | X |
| 219:35 | 223:37 | MSFN | X | | X | | | | |
| 221:06 | 221:11 | MAD | | | | | X | | |
| 223:37 | 227:32 | MSFN | X | | X | | | | |
| 223:40 | 225:40 | MAD | | | | | | | X |

[a]Data dump

[b]Indicates wing site.

TABLE D-II.- LUNAR MODULE DATA AVAILABILITY

| Time, hr:min | | Range station | Bandpass plots or tabs | Bilevels | Computer words | O'graph records | Brush records | Special plots or tabs | Special programs |
|---|---|---|---|---|---|---|---|---|---|
| From | To | | | | | | | | |
| -04:00 | 00:00 | ALDS | X | | | | | | |
| +07:50 | +08:00 | MSFN | X | | | | | | |
| 89:58 | 90:20 | HSKX[b] | X | | | | X | | |
| 104:03 | 105:00 | GDS | | X | | | | | |
| 104:05 | 106:38 | MSFN | X | | X | | | | |
| 105:46 | 106:04 | GDS | | X | | X | | | X |
| 106:03 | 106:40 | GDS | | X | | | | | |
| 106:38 | 111:20 | MSFN | X | | X | | | | |
| 106:40 | 106:59 | GDS | | X | | | | | |
| 107:46 | 108:33 | GDS | X | X | X | X | | | |
| 108:32 | 108:57 | GDS | X | X | | | | | |
| 108:57 | 109:25 | GDS[a] | X | X | | | | | |
| 108:58 | 110:34 | MSFN | | | X | | | | |
| 109:22 | 109:25 | GDS | | | | | | X | |
| 110:10 | 110:46 | GDS | X | X | X | X | X | X | X |
| 110:46 | 111:52 | GDS | | X | X | | | | |
| 110:20 | 115:39 | MSFN | X | | X | | | | |
| 111:50 | 113:02 | GDS | | X | X | | | | |
| 113:02 | 115:42 | HSK | | X | | | | | |
| 115:41 | 118:57 | MSFN | X | | | | | | |
| 115:44 | 119:33 | HSK | | X | | | | | |
| 119:17 | 123:06 | MSFN | X | | | | | | |
| 119:20 | 119:30 | HSK | | | | | X | | |
| 119:22 | 123:26 | MAD | | X | | | | | |
| 123:06 | 127:40 | MSFN | X | | | | | | |
| 123:26 | 128:27 | MAD | | X | | | | | |
| 127:41 | 131:44 | MSFN | X | | | | | | |
| 128:27 | 129:33 | MAD | | X | | | | | |
| 129:33 | 132:44 | GDS | | X | | | | | |
| 131:44 | 135:39 | MSFN | X | | | | | | |
| 131:45 | 135:58 | GDS | | X | | | | | |
| 135:39 | 139:20 | MSFN | X | | | | | | |
| 136:08 | 139:33 | HSK | | X | | | | | |
| 139:31 | 143:30 | MSFN | X | | X | | | | |
| 139:33 | 141:52 | HSK | | X | | X | | | |
| 141:52 | 142:21 | HSK | X | X | X | X | X | X | X |
| 142:19 | 142:32 | HSK | X | X | | X | | | X |
| 142:30 | 143:11 | HSK | | X | | | | | X |
| 143:11 | 143:52 | MAD | | X | | | | | |
| 143:40 | 147:28 | MSFN | X | | X | | | | |
| 143:44 | 144:05 | HSK | | X | X | X | | | X |
| 144:04 | 144:30 | MAD | | X | | | | X | |
| 145:11 | 145:50 | MAD | | X | X | | | X | |
| 145:50 | 147:39 | MAD | | X | | | | | |
| 147:28 | 150:06 | MSFN | X | | X | | | | |
| 147:39 | 149:56 | MAD | X | X | X | X | | X | X |

[a]Data dump

[b]Indicates wing site.

## APPENDIX E - MISSION REPORT SUPPLEMENTS

Table E-I contains a listing of all supplemental reports that are or will be published for the Apollo 7 through Apollo 12 mission reports. Also indicated in the table is the present status of each report not published or the publication date for those which have been completed.

## TABLE E-I.- MISSION REPORT SUPPLEMENTS

| Mission | Supplement number | Supplement title | Publication date /status |
|---|---|---|---|
| Apollo 7 | 1 | Trajectory Reconstruction and Analysis | May 1969 |
| Apollo 7 | 2 | Communications System Performance | June 1969 |
| Apollo 7 | 3 | Guidance, Navigation, and Control System Performance Analysis | November 1969 |
| Apollo 7 | 4 | Reaction Control System Performance | August 1969 |
| Apollo 7 | 5 | Cancelled | |
| Apollo 7 | 6 | Entry Postflight Analysis | December 1969 |
| Apollo 8 | 1 | Trajectory Reconstruction and Analysis | December 1969 |
| Apollo 8 | 2 | Guidance, Navigation and Control System Performance Analysis | November 1969 |
| Apollo 8 | 3 | Performance of Command and Service Module Reaction Control System | Final review |
| Apollo 8 | 4 | Service Propulsion System Final Flight Evaluation | Final review |
| Apollo 8 | 5 | Cancelled | |
| Apollo 8 | 6 | Analysis of Apollo 8 Photography and Visual Observations | December 1969 |
| Apollo 8 | 7 | Entry Postflight Analysis | December 1969 |
| Apollo 9 | 1 | Trajectory Reconstruction and Analysis | November 1969 |
| Apollo 9 | 2 | Command and Service Module Guidance, Navigation, and Control System Performance Analysis | November 1969 |
| Apollo 9 | 3 | Lunar Module Abort Guidance System Performance Analysis | November !969 |
| Apollo 9 | 4 | Performance of Command and Service Module Reaction Control System | Final review |
| Apollo 9 | 5 | Service Propulsion System Final Flight Evaluation | December 1969 |
| Apollo 9 | 6 | Performance of Lunar Module Reaction Control System | Preparation |
| Apollo 9 | 7 | Ascent Propulsion System Final Flight Evaluation | December 1969 |
| Apollo 9 | 8 | Descent Propulsion System Final Flight Evaluation | Preparation |
| Apollo 9 | 9 | Cancelled | |
| Apollo 9 | 10 | Stroking Test Analysis | December 1969 |
| Apollo 9 | 11 | Communications System Performance | December 1969 |
| Apollo 9 | 12 | Entry Postflight Analysis | December 1969 |
| Apollo 10 | 1 | Trajectory Reconstruction and Analysis | Final review |
| Apollo 10 | 2 | Guidance, Navigation and Control System Performance Analysis | December 1969 |
| Apollo 10 | 3 | Performance of Command and Service Module Reaction Control System | Final review |
| Apollo 10 | 4 | Service Propulsion System Final Flight Evaluation | Rework |
| Apollo 10 | 5 | Performance of Lunar Module Reaction Control System | Preparation |
| Apollo 10 | 6 | Ascent Propulsion System Final Flight Evaluation | January 1970 |
| Apollo 10 | 7 | Descent Propulsion System Final Evaluation | January 1970 |
| Apollo 10 | 8 | Cancelled | |
| Apollo 10 | 9 | Analysis of Apollo 10 Photography and Visual Observations | Preparation |
| Apollo 10 | 11 | Communications Systems Performance | December 1969 |
| Apollo 10 | 11 | Entry Postflight Analysis | December 1969 |
| Apollo 11 | 1 | Trajectory Reconstruction and Analysis | Preparation |
| Apollo 11 | 2 | Guidance, Navigation and Control System Performance Analysis | Final review |
| Apollo 11 | 3 | Performance of Command and Service Module Reaction Control System | Preparation |
| Apollo 11 | 4 | Service Propulsion System Final Flight Evaluation | Preparation |

| | | | |
|---|---|---|---|
| Apollo 11 | 5 | Performance of Lunar Module Reaction Control System | Preparation |
| Apollo 11 | 6 | Ascent Propulsion System Final Flight Evaluation | Preparation |
| Apollo 11 | 7 | Descent Propulsion System Final Flight Evaluation | Preparation |
| Apollo 11 | 8 | Cancelled | |
| Apollo 11 | 9 | Apollo 11 Preliminary Science Report | December 1969 |
| Apollo 11 | 10 | Communications Systems Performance | January 1970 |
| Apollo 11 | 11 | Entry Postflight Analysis | Preparation |
| Apollo 12 | 1 | Trajectory Reconstruction and Analysis | Preparation |
| Apollo 12 | 2 | Guidance, Navigation and Control System Performance Analysis | Review |
| Apollo 12 | 3 | Service Propulsion System Final Flight Evaluation | Preparation |
| Apollo 12 | 4 | Ascent Propulsion System Final Flight Evaluation | Preparation |
| Apollo 12 | 5 | Descent Propulsion System Final Flight Evaluation | Preparation |
| Apollo 12 | 6 | Apollo 12 Preliminary Science Report | Preparation |
| Apollo 12 | 7 | Landing Site Selection Processes | Preparation |

## APPENDIX F - GLOSSARY

| | |
|---|---|
| albedo | percentage of incoming radiation that is reflected by a natural surface |
| anorthositic | pertaining to a plutonic (originating far below the surface) rock composed almost wholly of plagioclase |
| basalt | generally, any fine-grained dark-colored igneous rock |
| breccia | a rock consisting of sharp fragments embedded in any fine grained matrix |
| ejecta | material thrown out as from a volcano |
| fayalitic | pertaining to a mineral consisting of an iron silicate isomeric (Fe2SiO4) with olivine |
| feldspar | any of a group of white, nearly white, flesh-red, bluish, or greenish minerals that are aluminum silicates with potassium, sodium, calcium, or barium |
| fines | very small particles in a mixture of sizes |
| gabbro | a medium- or coarse-grained basic igneous rock-forming intrusive body of medium or large size and consisting chiefly of plagioclase and pyroxene |
| hydrous | relating to water |
| igneous | formed by solidification from a molten or partially molten state |
| ilmenite | a usually massive, iron-black mineral of sub-metallic luster |
| induration | hardening |
| mafic | of or relating to a group of minerals characterized by magnesium and iron and usually by their dark color |
| modal | most common |
| morphology | study of form and structure in physical geography |
| olivine | mineral; a magnesium-iron silicate commonly found in basic igneous rocks |
| orthoclase | a type of feldspar |
| pegmatitic | pertaining to a natural igneous rock formation consisting of a variety of granite that occurs in dikes or veins and usually characterized by extremely coarse structure |
| pigeonite | mineral consisting of pyroxene and rather low calcium, little or no aluminum or ferric iron, and less ferrous iron than magnesium |
| plagioclase | a type of feldspar |
| polymorph | rock crystallizing with two or more different structures |
| pyroxene | a family of important rock-forming silicates |
| ray | any of the bright, whitish lines seen on the moon as extending radially from impact craters |
| regolith | fine grained material on the lunar surface |
| sanidine | a variety of orthoclase in often transparent crystals in eruptive rock, sometimes called glassy feldspar |
| scoria | rough, vesicular, cindery, usually dark lava developed by the expansion of the enclosed gases in basaltic magma |
| trachyte | a usually light-colored volcanic rock, consisting primarily of potash feldspar |

# The accompanying DVD -Video disc is Region 0.
# It is a double-sided disc.

## Apogee Books Space Series

| # | Title | ISBN | Bonus | US$ | UK£ | CN$ |
|---|---|---|---|---|---|---|
| 1 | Apollo 8 | 1-896522-66-1 | CDROM | $18.95 | £13.95 | $25.95 |
| 2 | Apollo 9 | 1-896522-51-3 | CDROM | $16.95 | £12.95 | $22.95 |
| 3 | Friendship 7 | 1-896522-60-2 | CDROM | $18.95 | £13.95 | $25.95 |
| 4 | Apollo 10 | 1-896522-52-1 | CDROM | $18.95 | £13.95 | $25.95 |
| 5 | Apollo 11 Vol 1 | 1-896522-53-X | CDROM | $18.95 | £13.95 | $25.95 |
| 6 | Apollo 11 Vol 2 | 1-896522-49-1 | CDROM | $15.95 | £10.95 | $20.95 |
| 7 | Apollo 12 | 1-896522-54-8 | CDROM | $18.95 | £13.95 | $25.95 |
| 8 | Gemini 6 | 1-896522-61-0 | CDROM | $18.95 | £13.95 | $25.95 |
| 9 | Apollo 13 | 1-896522-55-6 | CDROM | $18.95 | £13.95 | $25.95 |
| 10 | Mars | 1-896522-62-9 | CDROM | $23.95 | £18.95 | $31.95 |
| 11 | Apollo 7 | 1-896522-64-5 | CDROM | $18.95 | £13.95 | $25.95 |
| 12 | High Frontier | 1-896522-67-X | CDROM | $21.95 | £17.95 | $28.95 |
| 13 | X-15 | 1-896522-65-3 | CDROM | $23.95 | £18.95 | $31.95 |
| 14 | Apollo 14 | 1-896522-56-4 | CDROM | $18.95 | £15.95 | $25.95 |
| 15 | Freedom 7 | 1-896522-80-7 | CDROM | $18.95 | £15.95 | $25.95 |
| 16 | Space Shuttle STS 1-5 | 1-896522-69-6 | CDROM | $23.95 | £18.95 | $31.95 |
| 17 | Rocket Corp. Energia | 1-896522-81-5 | | $21.95 | £16.95 | $28.95 |
| 18 | Apollo 15 - Vol 1 | 1-896522-57-2 | CDROM | $19.95 | £15.95 | $27.95 |
| 19 | Arrows To The Moon | 1-896522-83-1 | | $21.95 | £17.95 | $28.95 |
| 20 | The Unbroken Chain | 1-896522-84-X | CDROM | $29.95 | £24.95 | $39.95 |
| 21 | Gemini 7 | 1-896522-80-7 | CDROM | $19.95 | £15.95 | $26.95 |
| 22 | Apollo 11 Vol 3 | 1-896522-85-8 | DVD* | $27.95 | £19.95 | $37.95 |
| 23 | Apollo 16 Vol 1 | 1-896522-58-0 | CDROM | $19.95 | £15.95 | $27.95 |
| 24 | Creating Space | 1-896522-86-6 | | $30.95 | £24.95 | $39.95 |
| 25 | Women Astronauts | 1-896522-87-4 | CDROM | $23.95 | £18.95 | $31.95 |
| 26 | On To Mars | 1-896522-90-4 | CDROM | $21.95 | £16.95 | $29.95 |
| 27 | Conquest of Space | 1-896522-92-0 | | $23.95 | £19.95 | $32.95 |
| 28 | Lost Spacecraft | 1-896522-88-2 | | $30.95 | £24.95 | $39.95 |
| 29 | Apollo 17 Vol 1 | 1-896522-59-9 | CDROM | $19.95 | £15.95 | $27.95 |
| 30 | Virtual Apollo | 1-896522-94-7 | | $19.95 | £14.95 | $26.95 |
| 31 | Apollo EECOM | 1-896522-96-3 | | $29.95 | £23.95 | $37.95 |
| 32 | Visions of Future Space | 1-896522-93-9 | CDROM | $27.95 | £21.95 | $35.95 |
| 33 | Space Trivia | 1-896522-98-X | | $19.95 | £14.95 | $26.95 |
| 34 | Interstellar Spacecraft | 1-896522-99-8 | | $24.95 | £18.95 | $30.95 |
| 35 | Dyna-Soar | 1-896522-95-5 | DVD* | $32.95 | £23.95 | $42.95 |
| 36 | The Rocket Team | 1-894959-00-0 | DVD* | $34.95 | £24.95 | $44.95 |
| 37 | Sigma 7 | 1-894959-01-9 | CDROM | $19.95 | £15.95 | $27.95 |
| 38 | Women Of Space | 1-894959-03-5 | CDROM | $22.95 | £17.95 | $30.95 |
| 39 | Columbia Accident Rpt | 1-894959-06-X | CDROM | $25.95 | £19.95 | $33.95 |
| 40 | Gemini 12 | 1-894959-04-3 | CDROM | $19.95 | £15.95 | $27.95 |
| 41 | The Simple Universe | 1-894959-11-6 | | $21.95 | £16.95 | $29.95 |
| 42 | New Moon Rising | 1-894959-12-4 | DVD* | $33.95 | £23.95 | $44.95 |
| 43 | Moonrush | 1-894959-10-8 | | $24.95 | £17.95 | $30.95 |
| 44 | Mars Volume 2 | 1-894959-05-1 | DVD* | $28.95 | £20.95 | $38.95 |
| 45 | Rocket Science | 1-894959-09-4 | | $20.95 | £15.95 | $28.95 |
| 46 | How NASA Learned | 1-894959-07-8 | | $25.95 | £18.95 | $35.95 |
| 47 | Virtual LM | 1-894959-14-0 | CDROM | $29.95 | £22.95 | $42.95 |
| 48 | Deep Space | 1-894959-15-9 | DVD* | $TBA | £TBA | $TBA |
| 49 | Space Tourism | 1-894959-08-6 | | $20.95 | £15.95 | $28.95 |
| 50 | Apollo 12 Volume 2 | 1-894959-16-7 | DVD* | $24.95 | £15.95 | $31.95 |

**CG Publishing Inc** home of **Apogee Books**
P.O Box 62034 Burlington, Ontario L7R 4K2, Canada
TEL. 1 905 637 5737 FAX 1 905 637 2631
e-mail marketing@cgpublishing.com
* NTSC Region 0

**Many more to come! Check our website for new titles.**

**www.apogeebooks.com**